HOW TO READ
BRIDGES

HOW TO READ
BRIDGES

A crash course in engineering and architecture

Rizzoli
NEW YORK

New York · Paris · London · Milan

Edward Denison
and Ian Stewart

First published in the United States
of America in 2012 by
Rizzoli International Publications, Inc.
300 Park Avenue South
New York, NY 10010
www.rizzoliusa.com

Copyright © 2012 Ivy Press Limited

ISBN: 978-0-7893-2491-7
2012 2013 2014 2015
10 9 8 7 6 5 4 3 2 1

Library of Congress Control Number:
2011938898

Color origination by Ivy Press Reprographics
Printed in China

This book was conceived, designed,
and produced by
Ivy Press
210 High Street
Lewes, East Sussex
BN7 2NS, UK
www.ivy-group.co.uk

CREATIVE DIRECTOR Peter Bridgewater
PUBLISHER Jason Hook
EDITORIAL DIRECTOR Caroline Earle
ART DIRECTOR Michael Whitehead
DESIGN JC Lanaway
ILLUSTRATOR Adam Hook
PROJECT EDITOR Jamie Pumfrey
PICTURE MANAGER Katie Greenwood

CONTENTS

INTROD

"Necessity," so the saying goes, "is the mother of invention." For bridge building, this truism is the ultimate raison d'être. From the earliest times when humans needed simply to cross a stream to reach new pastures, to the present day when global trade and communication depend on myriad bridges and complex bridge systems crossing rivers and seas, the need for new bridges has been a constant source of innovation.

Bridges have played a crucial role in humankind's expansion around the globe. From humble origins when bridges were built to fulfill the primitive needs of our ancestors, bridges have helped to define the evolution of settlements, towns, cities, and even nations. The outcome of territorial disputes has often hinged on the control of bridges. Throughout history the development of commerce has relied on dependable river crossings. In modern times, the expansion of cities and sometimes even the relationship between countries owe a debt to bridges. It is impossible to imagine the world's largest cities, such as New York or London, without their river crossings, just as ancient cities, such as Venice or Rome, could not have prospered without their bridges.

In recent decades, bridge design and construction has witnessed a prodigious growth. The records for the world's longest, tallest, highest, busiest, and heaviest bridges have been repeatedly broken as our knowledge of engineering, materials technologies, and construction skills has improved and is constantly satisfying our incessant demand for better means of communication. Today, the latest innovations in design and engineering

UCTION

are employed to construct new bridges big enough to link nations across seas and small enough to create new links and exciting routes in historic cities.

For the engineer, bridges are one of the most stimulating and explicit manifestations of their trade. For the rest of us, most bridges are taken for granted. While some bridges might inspire awe because of their size or form and others might evoke sentimentality because of their age, most bridges are seen with little more reverence than being merely useful or utilitarian. However, being able to read a bridge brings all these structures to life, from the most celebrated examples to the apparently innocuous.

In this book, reading bridges is arranged into two parts. The first part begins with the basics, examining the materials that have been used to build bridges throughout history, reviewing

the many different types of bridges, understanding the variety of uses, and introducing some of the most renowned engineers whose names have become synonymous with bridge design. The second part explores a wide selection of bridges from across the globe through a series of case studies arranged according to bridge type. Each case study helps the reader understand the structure and interpret its distinguishing characteristics.

Understanding why a structure looks the way it does why it is built the way it is, and why it uses the materials it does excites the inquisitive mind and opens the untrained eye to a unique field of human ingenuity that can be found all around us. From the busiest city center to the remotest mountain pass, every bridge has its own story—as long as the viewer is able to read it.

Looking for Clues

Structural form
A siding of stone was originally planned for the towers of the George Washington Bridge in New York. However, they were left uncovered, since the exposed structural form was considered more attractive and more economical.

In the modern built environment, bridges are almost unique in presenting an exposed structural form. Old stone or masonry buildings, with their buttresses, arches, and piers, offered a visual honesty that enabled interested observers to appreciate the structure and develop an intuitive understanding of how they stood up. However, as modern building materials have developed, steel, concrete, and wooden structures are commonly enshrouded in siding materials that create the aesthetic character of a building while concealing its primary structure.

Bridges similarly have developed rapidly with the material advances of the last two centuries but generally have no need for siding. These structures, therefore, remain one of the truest combinations of engineering and architecture, presenting a great opportunity to observe and understand structural form.

Direction of bracing

Simple geometric changes can significantly alter the way a structure transfers load back to its supports. In beam-truss configurations, if the bracing slopes down toward the center of the span, it carries load in tension. If it slopes upward, then it works in compression and must, therefore, be bigger in cross section to withstand buckling forces. However, cantilever configurations can reverse these rules, as shown above.

Joints

Look for joints. These can indicate simply supported beams instead of continuous beams, and cantilever bridges instead of beam or arch bridges. They can also reveal how the bridge was constructed, often with sections fabricated off-site.

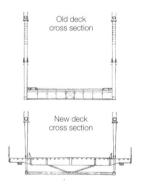

Old deck
cross section

New deck
cross section

Hidden foundations

Closer inspection of different arch types can give an indication of how the foundations are working. Deck-arch bridges require foundations that provide resistance to both vertical and lateral components of thrust forces. A tied-arch bridge uses the deck to develop tensile forces, meaning that little or no lateral force reaches the foundations. By observing and understanding the superstructure, we can often tell a great deal about foundations that we never get to see.

Authenticity

Maintenance is an important consideration for all bridges. Components have to be replaced or repaired over the working life of the structure. Similarly, many old bridges have had to adapt to changes in usage and increased demands on their capabilities. Some have had their decks modernized, while others are rebuilt to maintain the knowledge of their construction. Look for nonoriginal elements that give clues about their evolution.

Future Bridges

Although many of the fundamental concepts of bridge design remain the same as they have done for thousands of years, the past two centuries have witnessed a revolution in bridge building. New materials and technologies, and the accumulative experience and improving expertise of designers and engineers, have helped to create longer, more efficient, and safer structures than ever before.

Bridge building has enjoyed a period of growth that is set to continue into the 21st century. In recent years, world records for the longest, tallest, and highest bridges across all categories of type or use have been repeatedly broken. The escalation of record-breaking bridge building is caused in part by the refinement of expertise and the

technological tools available to designers and engineers, and in part by the needs of developing nations as they build modern infrastructure. Globalization and the political and economic union of nations and regions will also continue to fuel bridge building.

The future of bridge building is one of unparalleled potential, as designers and engineers continue to push the boundaries of established knowledge. In the short term, the refinement of conventional techniques will create longer and more efficient structures, while in the long term new materials and technologies will inevitably generate entirely new structural types that will in turn revolutionize bridge design, just as steel and concrete have done in our time.

Third Forth Bridge, Scotland
This new bridge is designed as a composite cable-stayed and continuous-beam structure. The road deck will be carried across the estuary by a combination of three A-frame, cable-stayed pylons and concrete piers. The cable-stayed section will create two elevated main navigation channels and two side spans, and the piers will support the approach roads that gently incline from the shoreline to the cable-stayed section. The A-frame pylons will support two sets of cables on each side, arranged in a fanlike pattern. These cables will be connected to the sides of the road deck.

Future Bridges

Messina Strait Bridge

The idea of bridging the Messina Strait that separates mainland Italy from Sicily goes back as far as the Romans and has not dwindled since. Serious but ultimately fruitless studies were conducted in the late 19th and early 20th centuries, and then renewed interest in the idea in the 1950s led to a competition being held in the 1960s, but also with no success. The project finally received state approval in the 1990s, and in 2006 a plan was drafted for the longest and tallest bridge structure ever built. The project faced cancellation but was reinstated in 2009. Although its immediate future remains in doubt, the bridging of the Messina Strait is an idea that will not go away, and when finally realized, it will represent a milestone in the history of bridge building.

Historic future bridge

The proposed Messina Strait suspension bridge has a single span of 10,800 feet, 60 percent longer than the current record holder, the Akashi Kaikyo Bridge (see pages 218–219). To provide a sufficient gradient for the suspension cables and a 213-foot clearance for shipping, the pylons have to be almost 1,312 feet high, exceeding the record holder, the Millau Viaduct (see pages 230–231), by more than 130 feet. Two pairs of 5-foot-diameter cable are required to carry the 197-foot-wide deck, comprising six lanes of traffic, two railroad lines, two pedestrian lanes, and an emergency lane.

The Third Forth Bridge

The Firth of Forth in Scotland is accustomed to record-breaking bridge design. The first bridge across the Forth was the rail bridge (see page 181). Opened in 1890, it was the world's longest cantilever structure and the first large bridge made entirely of steel. The second bridge was the longest suspension bridge outside the United States when it opened to vehicular traffic in 1964 (see pages 35 and 56). However, the deterioration of its main cables over half a century has weakened the structure and a new bridge has been designed to replace it. Construction is due to be completed in 2016.

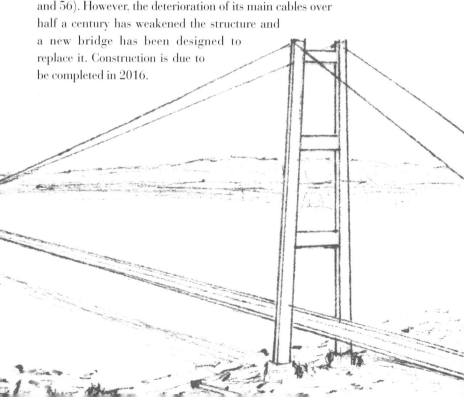

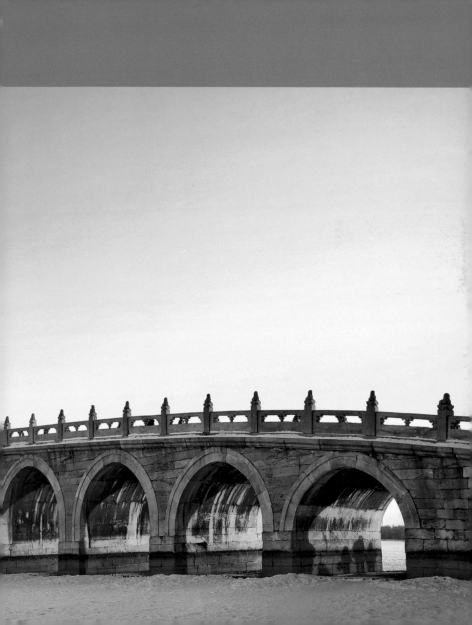

MATERIALS Introduction

Throughout history and across the globe, bridge building has relied on an amazing array of materials. Some, such as wood and stone, have endured for millennia, while others, such as bamboo, roots, or vines, are unique to their place. Other materials require specialized manufacturing. Kiln-fired bricks, dating back more than four thousand years, were among the first examples of a manufactured material that was later used for bridge construction.

Over the last two centuries, modern materials and manufacturing processes have revolutionized bridge building. Iron, steel, concrete, and even toughened glass have transformed bridge design and enhanced our capacity to bridge longer distances and design more innovative, efficient, and safer structures.

It is not just the primary materials from which bridges are built but also the secondary materials that have contributed to their evolution. Lime mortar was used to bond stones, while iron nails strengthened wooden bridges and iron dovetail joints tied stones together. Most modern bridges rely on the combined and complementary attributes of concrete and steel. Whether in reinforced concrete, prestressed concrete, or posttensioned concrete, steel's tensile properties augment concrete's compressive qualities.

The advancement of bridge design ultimately relies on a material to perform safely, effectively, and efficiently. With materials technologies and the development of composite materials constantly improving, the future potential in bridge design is almost boundless.

Enduring wood
The Chengyang Wind and Rain Bridge in Guangxi Province, China, was built in 1916. The elaborate deck, with its five pavilions and numerous terraces, is constructed of wood; the three piers are built in stone; and the roof is covered in ceramic tiles.

Stone

Stone is an ancient bridge-building material and the most enduring. Although it is weaker under tension than compression, stone's durability and strength make it particularly suited to bridges. Stone bridges take advantage of stone's compressive strength by arranging the material vertically (as piers) and as arches. Some bridges rely on the tensile strength of stones, such as ancient or very small bridges, where a slab is placed across a small stream or series of pillars traversing a river.

Bridging communities
The old stone bridge (Stari Most), in Mostar, Bosnia and Herzegovina, a 13-foot-wide and 66-foot-high humpbacked-arch bridge, was built in 1566 by Mimar Hajruddin, a pupil of one of the Ottoman Empire's most famous architects, Sinan. It bridged communities on both sides of the Neretva River for centuries before its destruction in 1993 during the Croat–Bosniak War. It was faithfully reconstructed and reopened in 2004.

Natural stone arches (right)

The oldest stone bridges occur naturally as eroded arches on coastlines and exposed areas of bedrock around the world. In Utah, the Arches National Park contains more than two thousand of these natural wonders, the most famous of which is the 52-foot-high Delicate Arch. These natural formations would have inspired ancient civilizations and demonstrated the capability of stone.

Stone beams (left)

The simplest stone bridges use stone as a beam, relying on its tensile qualities. As the distance between the supports at each end increases, so must the thickness of the stone. Although simple in theory, the technical sophistication of stone-beam bridges varies from the Chinese bridges, where individual stones can weigh more than 200 tons to the diminutive clapper bridges in Britain.

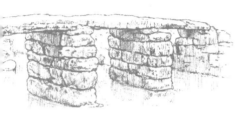

Stone ornament

The sculptural quality of stone is exploited in ornamental bridges, such as the Roman triumphal bridges, which performed important decorative or ceremonial functions and were furnished with elaborately carved stonework.

Wood

MATERIALS

Flying eaves

The Chengyang Wind and Rain Bridge in China's Guangxi Province was built entirely of wood in 1916. It uses a method of construction known as flying eaves, whereby layers of overlapping beams support a load in the same manner as in the distinctive traditional Chinese roof. Multistoried pavilions stand on the five stone piers supporting each section of this 210-foot-long and 10-foot-wide covered pedestrian bridge.

Wood is an ideal material for bridge building—it is abundant, lightweight, comparatively cheap, easy to work, and extremely versatile. Consequently, wooden bridges are found all over the world and come in all shapes and sizes, from the ornamental elegance of dainty footbridges to the utilitarian solidity of massive railroad bridges. Wood's structural dexterity allows for it to be used to carry loads in compression or in tension and also enables it to be easily crafted into intricate shapes and designs.

Wooden trusses (right)

The first American covered wooden bridge was built across the Schuylkill River, in Pennsylvania, in 1805. Depicting the bridge before it was covered, this engraving reveals how the deck is fabricated from wooden trusses with arched profiles that span between two stone piers.

Wooden arches

The five slender arches (each approximately 115 feet long and 16 feet wide) of the Kintaikyo Bridge that leap across the Nishiki River in Japan were built in 1673. The complex structure, originally built entirely of wood, is designed to resist heavy flooding. It was last rebuilt in 1953.

Ancient traditions

The 322-foot-long Wan'an Bridge in Fujian, China, built more than a thousand years ago, is a testament to wood's potential endurance. The region is famous for its covered wooden footbridges, each of which is constructed from segmental arches made up of wooden beams supported on stone piers.

Organic Material

Modern vines
Today, there are
three bridges across
the remote Iya
Valley of southern
Japan measuring
up to 147 feet long
and 6½ feet wide.

One of the earliest methods of bridging large rivers was by felling a tree across the banks. This had its limitations—for example, if a gorge was too wide or if there were no suitable trees. In some regions, nature offers assistance in other ways, such as with vines and vegetation hanging from the forest canopy. Over time, more sophisticated ways of manipulating organic materials other than wood into bridges have been invented.

Vines (right)

For centuries, the only means of crossing the Iya Valley was the vine bridges, or *kazurabashi*, suspended 46 feet above the river. These were first constructed in the 12th century so that retreating warriors could easily cut the vines to prevent their enemy's advance. Each one is regularly reconstructed, but all are now reinforced with steel wires.

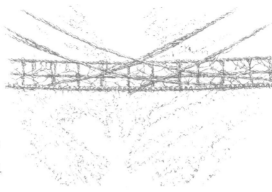

Roots (left)

In the drenched jungles of northeast India, the locals have devised a unique way of constructing bridges from a native fig tree, *Ficus elastica*. Instead of cutting the tree down, they manipulate the aerial roots growing from the tree's branches and train them across the river to create the suspension cables of living bridges that grow stronger with age. The decks are paved with stones and sometimes built over two stories, creating double-deck suspension bridges.

Bamboo

Bamboo has been used in Asia for bridge construction for thousands of years. Its high compressive and tensile strength, flexibility, and lightness make it ideal for floating or suspension bridges carrying relatively small loads. The longest bamboo suspension bridges, up to 655 feet long, were constructed in China's Sichuan Province. As a processed material, bamboo is now being used to build bridges capable of carrying vehicular traffic.

Brick

Ancient brick

The first sun-dried bricks were manufactured in Mesopotamia six thousand years ago. In the 17th century, their modern equivalent, kiln-fired bricks, were used to construct the thirty-three-arch Si-o-se Pol Bridge in nearby modern-day Iran, one of the oldest and longest brick bridges in the world. Constructed over the Zayandeh River in the city of Isfahan, the 965-foot-long bridge has a central stone deck for vehicles that is flanked by covered pedestrian walkways, which are separated by walls of sixty-six arches (two above each lower arch).

Bricks have been fabricated for thousands of years and are one of the earliest examples of a manufactured material used in bridge construction. In warm climates, the earliest bricks were made from uniform blocks of mud, clay, or similar base materials and left to dry. The manufacture of weatherproof bricks relies on specialty materials and the expertise to burn bricks in kilns, which was first recorded in China's ancient capital, Xian, nearly four thousand years ago. Centuries later, the Romans used similar techniques, which, following their downfall, were lost in northern Europe until the Late Middle Ages.

Massive structures

The Goeltzschtalbruecke, or Göltzsch Viaduct, of southeastern Germany is one of the most outstanding articulations of brick's versatility. Stretching over 1,640 feet and nearly 330 feet tall at its highest point, the multiarched viaduct opened in 1851 and continues to carry modern high-speed trains.

Reconstructed brick

The ancient twenty-two-arch brick bridge of Pul-i-Malan outside the city of Herat, in Afghanistan, is believed to date from the 12th century. After surviving many floods, its bricks were lauded in local legends, but they succumbed to modern warfare and the bridge has recently been substantially reconstructed.

Brick styling

Brick was a favored material of the Victorian era in Britain, and few architects used it more splendidly than Sir George Gilbert Scott, who designed the Clifton Hampden Bridge in Oxfordshire (1867). Scott's preference for the Gothic style of architecture, exemplified by his design for the Midland Hotel that fronts London's St. Pancras Station, appears in the bridge's six pointed arches.

Iron

The Chinese were the first to use iron in bridges, fabricating iron links to create suspension chains. Local historians documented them in the 15th century and subsequent European travelers, including Marco Polo, marveled at the sophistication of these pioneering suspension bridges. No other application of iron in bridge building matched these structures until the late 18th century, when the world's first cast-iron bridge was constructed in Coalbrookdale, England. Throughout the Industrial Revolution, iron was a popular construction material until it was superseded by steel.

Cast iron (right)

Iron transformed bridge construction during the 19th century. Opened in 1805, the 997-foot-long, 11-foot-wide, and 125-foot-high Pontcysyllte Aqueduct was designed by Victorian engineer Thomas Telford. The trough, made from cast-iron plates, is supported on eighteen arches, each formed of four cast-iron ribs, and reinforcing outer plates bridge each of the masonry piers.

Iron Bridge (left)

No iron bridge is more famous than the one built across the Severn River in England, whose very name, Iron Bridge, derives from the material. The revolutionary bridge, which was a resounding advertisement for iron, was designed by Thomas Farnolls Pritchard using about 417 tons of iron produced continuously over three months from nearby furnaces. Each huge cast-iron member was transported to the site along the river and then hoisted into place to form an arched, humpbacked bridge that opened on the first day of 1781.

Decorative iron

Industrialization made it easier and cheaper to manufacture identical structural members that were not only structurally effective but could also be highly decorative. Elaborate shapes and patterns could be created during the casting process, and they swiftly became a feature of iron and steel bridges.

Steel

Bridging the Tyne
More than 7,700 tons of steel were used to construct the Tyne Bridge linking the cities of Newcastle and Gateshead. It was designed by Mott, Hay and Anderson. The steel arch, fabricated from flat sheets of steel riveted together from which the road deck is hung, has a 528-foot span. It was the longest single-span bridge in Britain when opened by King George V in 1928.

The advent of modern steelmaking marked a significant milestone in the development of bridge building in the late 19th century. Steel's superior strength in tension and compression compared with iron advanced the performance and construction techniques of metal bridges. Extruded steel wire was used to build the first modern suspension bridges, such as the Brooklyn Bridge in 1883, while other types of record-breaking bridges, including the ⅔-mile-long Forth Rail Bridge, the world's longest cantilevered bridge, were built from thousands of individual members joined together on-site.

Steel arch

With a span of more than 1,640 feet and a road deck 165 feet above the waters of Kill Van Kull, New Jersey, the Bayonne Bridge, completed in 1931, was the largest steel arch bridge in the world until 1978. The arch is made up of forty straight steel segments. Unlike similar bridges, it does not have decorative stone abutments.

Steel truss

The Brinnington Rail Bridge in Manchester, England, is a modern steel-truss bridge that demonstrates the enduring quality of steel as a popular bridge-building material. The individual truss sections are riveted to prefabricated steel girders that form the upper chord and deck of the bridge.

Concrete

Prefabrication
An advantage
of prestressed
concrete is that it
can reduce building
times and costs by
being prefabricated
in sections and
delivered to the
construction site.
By using prestressed
concrete, the new
Victory Bridge
(2005) in New
Jersey, whose
440-foot main
span made it the
longest precast
cantilever bridge in
the United States,
saved more than a
year's construction
time and millions
of dollars.

In 1849, a French gardener, Joseph Monier, stumbled upon a radical invention when using iron mesh in concrete to make flowerpots. By combining the tensile strength of iron and the compressive strength of concrete, ferroconcrete, commonly known as reinforced concrete, has since revolutionized the construction industry. Reinforced concrete has limited deflection and cracking characteristics, so prestressed concrete was invented in the early 20th century to enhance performance, thus enabling longer spans without increasing the depth of each section and, consequently, its weight.

Posttensioned concrete (right)

Posttensioning is similar in principle to prestressing. It is another method of increasing concrete's spanning capabilities by inducing compression forces in the parts of the structure likely to develop tensile forces. Unlike prestressing, posttensioning is carried out after the concrete has been formed and set. Steel tendons within casing sleeves are located in the member either before the concrete is poured in situ or, in the case of prefabricated elements, as the individual segments are lifted into place. They are then pulled tight, forcing the concrete into compression. The tendons are then locked off.

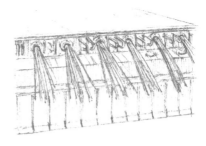

Slim line (below)

The structural qualities of prestressed concrete allow for large structures to be built using a comparatively small amount of material. The slender appearance of the 827-foot span arch across the Wilde Gera Valley, in Germany, would not have been achievable using standard reinforced concrete.

Prestressed concrete

Prestressing concrete is a method of enhancing the material's spanning capabilities. The concrete is cast around pretensioned steel cables, called tendons. As the concrete cures, a bond is formed between the tendon and the concrete. After curing, the tendon ends are released, forcing the concrete into compression. The tendons are positioned where tensile forces develop, usually at the bottom portion of a simple span.

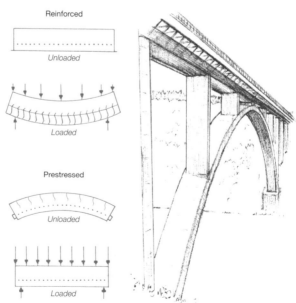

Reinforced

Unloaded

Loaded

Prestressed

Unloaded

Loaded

Glass

Revolutionary glass
A footbridge over
the Grand Union
Canal in London,
combines an
11½-foot-diameter
glass tube with
a 24½-foot-long
stainless-steel
helix to create a
functional bridge
and a work of art.
To let boats pass,
the helix rotates and
the bridge retracts.

Glass is among the most exhilarating materials to be used in bridge construction. Modern glassmaking techniques have allowed for glass to overcome many of the problems that until relatively recently prevented it from being used in bridges. Today, toughened glass can be manufactured in almost any shape and size. Glass is not only functional and safe, but it is also durable and resistant to heavy wear and weathering. Although glass has exceptional compressive strength, most applications in bridge design rely on other materials for structural integrity, such as steel frames. This will almost certainly change as the performance of glass improves, allowing for bridges to be built entirely of glass.

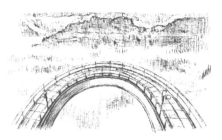

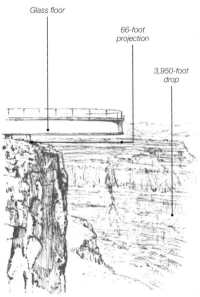

Glass floor

66-foot projection

3,950-foot drop

Transparency (above and right)

The obvious advantage of glass is its transparency. Few things match the thrill of being able to enjoy unobstructed views of scenery beneath your feet. Reinforced glass floors have been installed in high-rise structures for years, but nowhere in the world compares to the glass floor of the cantilevered horseshoe bridge over the Grand Canyon. With a drop of more than 3,950 feet, the glass-and-steel bridge creates the highest view from a man-made structure in the world.

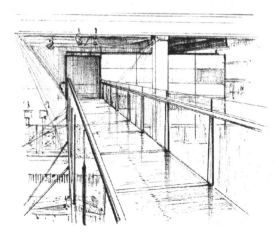

Interior design

Glass bridges are becoming a common feature of modern interior design, particularly in commercial or office spaces, and even in some public buildings. In 1997, a glass bridge suspended on 1/32-inch-thick steel wires was installed in the Challenge of Materials gallery of London's Science Museum.

Introduction

Irrespective of age or materials, every bridge can be categorized according to its type. The four basic types of bridges are beam, arch, cantilever, and suspension. The sequence of these types also reflects an approximate chronology of the development of bridge building. The most basic bridge type is the beam, which can be as simple as a log across a stream. Building an arch requires an element of constructional acumen, and a cantilever relies on a more developed appreciation of engineering. Although primitive examples of suspension bridges exist, modern suspension bridges offer the largest spans and are often the most challenging to build.

Other types of bridges are variations on these four basic designs. Cable-stayed bridges, for example, are a development of suspension and cantilever bridges, whereby each cable is attached to the tower and secured directly to the deck instead of carrying the loads from the deck through hangers to two main cables hung over the towers.

Truss and moving bridges, on the other hand, can employ any one (or a combination) of the four main types in a structural configuration that is either constructed from truss sections or is movable. Many bridges are designed using a combination of types, and such bridges are frequently referred to as hybrid.

Old and new
The Forth Road Bridge (1964) was the first bridge in Europe to employ the modern method of cable spinning during construction, using more than 18,640 miles of wire. Behind is the cantilevered Forth Rail Bridge (1890).

Beam Bridges

Box girders

One variation of the beam is the box girder, which gains its strength from a hollow, rectangular cross section. Prestressed-concrete box girders are used in the construction of the 8-mile-long and 236-foot-wide Rio–Niterói Bridge (1974) in Brazil.

In its simplest form a beam bridge is a horizontal deck supported at both ends, but there are many different types. A beam bridge has to withstand vertical loads that generate vertical shear forces, horizontal tension, and compression forces, and, when only loaded on one side, torsional twisting forces over its length. The vertical shear forces are shared between the supports. A beam's spanning capability can be intuitively characterized by the ratio between the depth of its cross section and the distance between the supports (the span-to-depth ratio)—if the distance between the supports at each end is too great, the beam may develop forces that it is not strong enough to withstand.

Basic beams

The simplest illustration of the beam bridge, and possibly the earliest example used by man, is the fallen tree across a river. Here, the beam comprises the tree trunk, which is supported at both ends by the river banks.

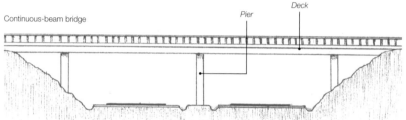

Continuous-beam bridge

Pier

Deck

Continuous beam (above)

An alternative to the simply supported span bridge is the continuous beam, which uses a single beam supported over numerous piers. The advantage of a continuous beam is that depths can be reduced because the beam works more efficiently by developing tension forces in the top portion of its cross section over the piers.

Simply supported beams (below)

Bridges that comprise individual beams between supports are known as simply supported span bridges. This means that although there may be repetition (numerous supports and numerous spans), the beam is not a continuous member over the top of the supports. Simply supported beams are used where the overall span requires only one length of beam or multiple spans where movement joints are required.

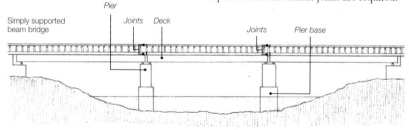

Pier

Simply supported
beam bridge

Joints   Deck

Joints   Pier base

Beam Bridges

Stone beams

Before modern advances in engineering and materials, beam bridges spanning great distances used a large number of beams supported on closely spaced piers.

Luoyang Bridge in Quanzhou, China, built in 1060 CE, is more than ⅔ mile long. Its forty-six masonry piers support huge carved granite stones.

Spreading the load (left)

Spreading the load across a series of parallel beams, as illustrated in this conventional highway bridge, helps to overcome twisting that could occur when weight is unevenly distributed on the deck. Here, the load on the deck is supported by four evenly spaced steel girders connected at each pier, which transfers the load into a single column, minimizing the bridge's impact at ground level.

Luoyang Bridge beams (right)

With some beams as long as 66 feet, the thickness they need to support even their own weight ensures each one weighs up to 220 tons—an extremely inefficient ratio. Because eight beams are required for each 16-foot-wide span, the Luoyang Bridge was constructed with more than 350 of these huge stone beams linking forty-six streamlined masonry piers.

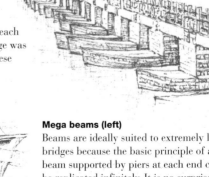

Mega beams (left)

Beams are ideally suited to extremely long bridges because the basic principle of a beam supported by piers at each end can be replicated infinitely. It is no surprise that the world's longest bridge—the Danyang–Kunshan Grand Bridge, 103 miles long, completed in 2010—is a beam bridge. Hundreds of prestressed-concrete beams have been laid across multiple piers.

Arch Bridges

BRIDGE TYPES

Line of forces
In an arch bridge, applied vertical forces are carried through the arch in compression and into the abutments, which resist these forces with vertical and horizontal reactions. In traditional stone arches, each block is called a voussoir and the stone at the apex is called the keystone. This locks the arch in place and ensures that vertical forces are translated into lateral lines of force.

The arch is naturally a very strong structure. Its occurrence in the natural environment predates and doubtlessly inspired human attempts to replicate it. These attempts were first in stone, but since then many other materials, including wood, brick, iron, steel, and prestressed concrete, have been used. An arch works by transferring the vertical load through its curve to the supports at each end, known as abutments, where the load is dissipated into the ground. There are several ways that bridges exploit the arch—some place the deck on top of the arch, some suspend the deck beneath the arch, and others allow for the deck to pass through the arch.

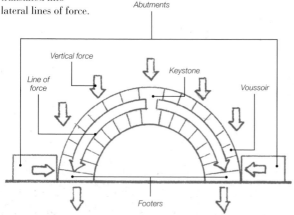

Abutments

Vertical force

Line of force

Keystone

Voussoir

Footers

Ancient arches

The simplest type of arch can be found in ancient stone structures, such as tombs, where two colossal stones placed diagonally against each other transfer the loads to each side of the opening beneath.

Half-through arches (below)

The Lupu Bridge (2003) is a box-steel arched bridge crossing the Huangpu River in Shanghai, China. With a span of 1,800 feet, it is the second longest arched bridge in the world. The bridge comprises two huge steel arches braced together, from which the deck of the bridge hangs on steel wires. Because the deck intersects the arch above the abutments, it is known as a half-through arch.

Arch Bridges

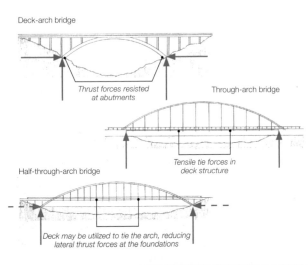

Deck-arch bridge

Thrust forces resisted at abutments

Through-arch bridge

Tensile tie forces in deck structure

Half-through-arch bridge

Deck may be utilized to tie the arch, reducing lateral thrust forces at the foundations

Arch types (left)

Single-span arched bridges traditionally caused traffic to climb up and over their humps. Nowadays, different types of arched bridges maintain a horizontal deck, including half-through arches, deck arches, and through arches, where the hanging deck intersects the arch at the abutment. (See also pages 110–111.)

Steel arch (right)

Modern construction materials and techniques have allowed for arch bridges to become increasingly slender. The lean lines of the half-through-arch bridge (1990) over Roosevelt Lake in Arizona is an example of how unobtrusive modern bridges can be when constructed in areas of outstanding natural beauty.

Open spandrel

Spandrels (above)

The inherent weight (deadweight) of any bridge can cause it to collapse, so reducing weight is often a key consideration in design. Creating holes in the space between the extrados, the exterior curve of an arch, and the abutment not only reduces the weight of the bridge, but it also lets floodwater pass through, reducing the bridge's resistance to water. These practical voids are called open spandrels, as seen above in the Pontypridd Bridge in southern Wales.

Deck arch (above)

An arch that supports the bridge deck above it is called a deck-arch bridge. The Eads Bridge over the Mississippi in St. Louis, which carries road and rail traffic, was the longest deck-arch bridge in the world when it opened in 1874. The load on a deck-arch bridge is supported by the arches beneath it and, in this case, also by the piers in between.

43

Truss Bridges

The classic truss
Many railroad bridges adhere to the classic image of a truss bridge. This bridge over the Colorado River illustrates the truss's structural geometry—triangular sections formed by the diagonal members between the vertical columns and the horizontal chords.

Truss bridges are one of the earliest forms of modern bridge design. There are many varieties of truss, but as a structure they all take advantage of the inherent strength of the triangle. The straight members in a truss are subject to compression, tension, or a combination of both (not at the same time), depending on the dynamic forces applied to the truss. Compression members are typically larger in cross section to withstand buckling effects. The truss is a very efficient structure because it uses comparatively little material and has a very good strength-to-weight ratio. Trusses can be used in all types of structures, including beams, arches, and cantilevers.

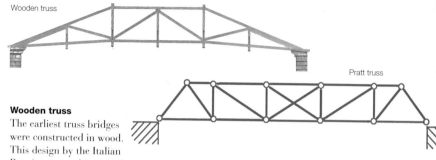

Wooden truss

Pratt truss

Wooden truss

The earliest truss bridges were constructed in wood. This design by the Italian Renaissance architect Andrea Palladio (1508–80) illustrates the truss's triangular geometry. The bridge is divided into three sections, with two inclining triangular approaches supported at one end by abutments and at the other end supporting the bridge's horizontal central section.

Pratt truss

This particular configuration, in which the diagonals slope down into the middle of the bridge, is called the Pratt truss and was invented by Thomas and Caleb Pratt in 1844.

Continuous truss

Some truss bridges are built using a repeated series of vertical, horizontal, and diagonal members to create a continuous structural framework. The Crumlin Viaduct (1857), in southern Wales, used continuous trusses in both the construction of the bridge deck and the supporting pylons. It was dismantled in 1967.

Truss Bridges

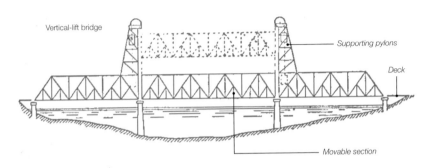

Vertical-lift bridge

Supporting pylons

Deck

Movable section

Moving loads (above)

The truss is well suited to moving bridges because it has a good resistance to moving loads. In the case of a lifting bridge, where heavy loads move across the bridge's span, the truss is capable of supporting the load at any position along its length, transferring it onto the supporting pylons on each side.

Hybrid trusses (below)

Often a variety of truss types are used in a single bridge. The original Kentucky and Indiana Terminal Bridge (1886) across the Ohio River used a number of different truss types, including arch, beam, and cantilever trusses constructed from wood and steel, throughout its ⅔-mile-long span.

Lenticular truss

So called because of the distinctive lens shape created by the upper and lower chords, the lenticular truss was used by Isambard Kingdom Brunel in his design of the Royal Albert Bridge (1859). Each truss comprises an upper chord of tubular iron in compression and two lower chords of iron chains in tension, which are cross braced to withstand wind load. The rail deck is suspended beneath. Unlike an arch, which transfers horizontal forces onto the piers, the lenticular truss is simply supported on the piers.

Lattice truss

The lattice truss uses a series of diagonal members laid in a lattice between larger and stronger horizontal members. This example shows how a truss can also be read as a beam, with each lattice-truss section forming a large beam supported on piers. Notice how the diagonals get larger as they carry the load back to the supports.

Moving Bridges

Rolling bascule
The most common
form of moving
bridge is the
bascule bridge,
which works on the
same principle as a
seesaw. The famous
Pegasus Bridge
(1934 and 1994)
in Normandy is
a rolling bascule—
instead of opening
at a pivot point
on the deck, the
structure rolls open.

Although bridges are designed to ease passage, they can also be an obstruction—or worse. A road bridge over water might be good for cars yet restrict boats, and a castle drawbridge that failed to withdraw would have been a matter of life and death. Moving bridges are designed to overcome the problems associated with the permanence of a bridge. There are a range of moving bridges to suit all kinds of situations, from bridges that rise into the air to ones that sink underwater, and from those that tilt to those that curl. As with all types of bridges, the design of moving bridges has evolved as new materials have become available.

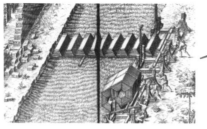

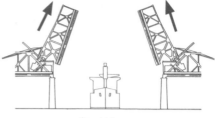

Completely open

Curling

Temporary bridges are often used in military situations where advancing or retreating soldiers can prevent their enemy from following them. They are lightweight and easy to move, like this 16th-century example that can be unfurled across a river or curled up to transport.

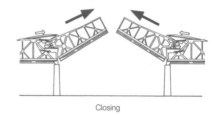

Closing

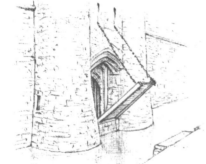

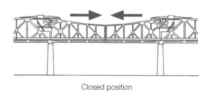

Closed position

Drawbridge

Among the most renowned moving bridges is the drawbridge, made famous by medieval European castles. The drawbridge opens and closes by using retractable ropes or metal chains attached to the farthest ends of the bridge's deck. Counterweights would often be used to make the process easier.

Bascule

The bridge deck, or "leaf," of a bascule bridge is raised using a counterweight, making the bascule energy efficient and swift. Bascule bridges can have either one leaf or two. A bascule bridge with two movable sections that meet in the middle is known as a double-leaf bascule. Lowering the counterweight located on the top of the truss section or on a concealed back span raises the leaf of the bridge.

Moving Bridges

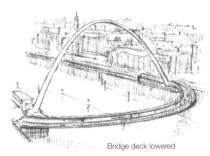

Bridge deck lowered

Bridge deck raised

Tilt (above)

Tilt bridges rotate at both ends, but for this action to have a useful purpose the bridge deck has to be curved, restricting its use to walking or cycling. In the case of the Millennium Bridge at Gateshead, the arch that supports the deck in its closed position rotates down as the deck lifts up, acting as a counterbalance. As the deck lifts, it is able to hold itself up without the arch.

Vertical lift (below)

Bridges that raise the central deck of the bridge vertically so that it remains parallel with the approaches are called vertical-lift bridges. The deck is raised either on two towers at each end of the span or by hydraulic jacks underneath the span. Vertical-lift bridges have a better weight-to-strength ratio than bascule bridges, so they can carrier heavier loads.

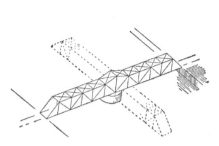

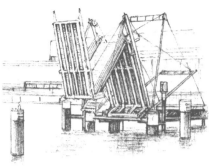

Swing

Swing bridges rotate from a central pivot in the middle of the span. Because they do not require a counterweight, they can be comparatively light, but the central pier from which they pivot can be a disadvantage and an obstruction in narrower channels.

Folding

Another way to retract bridges is to fold them like a concertina. Because of the comparatively fragile nature of moving parts, this method is suited to smaller bridges, where each section is not too heavy and the moving parts are not exposed to too much stress.

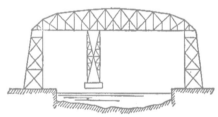

Transporter

Bridges that carry a section of deck across the span are known as transporter bridges. Most operate by suspending the deck on cables from an elevated structure supported on two piers at each end of the bridge. The deck, known as the gondola, moves from one side to the other using a mechanized system of steel cables and pulleys.

Submersible

The submersible bridge is a rare type of movable bridge that lowers into the water to let boats pass over the top. More of the boat sits above water than below, so a bridge that sinks into the water does not have to travel as far as one that rises.

Cantilever Bridges

Arch or cantilever?
Cantilever bridges can be mistaken for arched bridges because the two cantilever arms often form an arc, as they do here at Wandsworth Bridge, in London. The bridge comprises two cantilever sections standing on piers. The movement joints in the center of the span are just visible.

A cantilever is a structure projecting in one direction from an upright support. If the structure extends in both directions to form a T shape, it becomes a balanced cantilever. In bridge design, the two parts of the balanced structure are known as the cantilever arm and the anchor arm. Many different types of cantilever are used in bridge construction and they can often be difficult to identify, not least because they can masquerade as arches due to the arc shape of some cantilever arms. The easiest way to read a cantilever from an arch is to look for the joint between two cantilevers, which will be at or near the center of the span, whereas the joint between two arches will be at the pier.

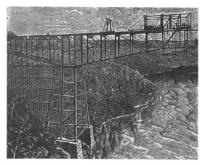

Niagara Cantilever Bridge during construction

Completed bridge

Niagara Cantilever construction

Constructing cantilever bridges is relatively straightforward because with the anchor arm tied back, the cantilever arm can be built out incrementally from the pier to form the main span. As a method, it is ideal to use over deep gorges or busy rivers because the bridge requires no temporary support from below. These illustrations of the construction and design of the bridge at Niagara Falls show the process of extending the cantilever arm out from the pier. The finished bridge comprises a pair of balanced cantilevers supporting a deck on top.

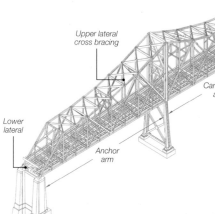

Lower floor beam

Upper lateral cross bracing

Suspended span

Cantilever arm

Lower lateral

Anchor arm

Reading the cantilever

Each section of a cantilever bridge can usually be read as three distinct parts: the anchor arm, the cantilever arm, and the suspended span. The pier attached to the anchor arm resists the uplift forces. The pier on which the cantilevers are balanced resists the compressive forces. The top chords are in tension. The bottom chords are in compression, and the diagonal members carry the load back to the support.

Cantilever Bridges

Oakland Bay Bridge

The east span of the double-deck Oakland Bay Bridge (1936), which is also often known as the Bay Bridge, crosses San Francisco Bay, California. The bridge comprises a series of small, medium, and large truss spans; the largest of these truss spans is 1,411 feet long and is formed by two large balanced cantilevers. The bridge is due to be replaced in 2013.

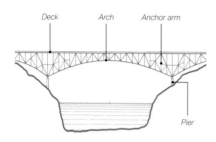

Deck　　　　Arch　　　Anchor arm

Pier

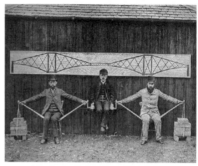

Balanced cantilever

The appearance of the arch formed on the underside of the two cantilever arms is misleading because the bridge does not function as an arch. Instead of transferring loads through the arch to abutments, the loads in a balanced-cantilever structure are transferred to the piers. The downward forces in the main span are countered by the anchor arm, which is held securely to the piers of the approach road.

Extending the span

The central span of cantilever bridges is often lengthened by inserting a suspended section between each cantilever arm. In this classic photograph, the person seated in the middle represents the suspended span supported by the cantilevers on each side. The wooden posts represent the lower chords in compression and the men's arms represent the upper chords in tension. Both sets of chords are anchored at each end to prevent the cantilevers from toppling inward.

Cantilever-truss span

Cantilever truss

Cantilever bridges are commonly truss bridges. Making the structure stronger over the supports causes it to behave as a cantilever, with the central section acting as a suspended beam. Placing two balanced cantilevers next to each other creates a large span between their supporting piers. The Oakland Bay Bridge is a good example of this method of cantilever-bridge design, in which the cantilever arm extends out to form half the main span and is secured by the anchor arm, which is attached to piers.

Suspension Bridges

The cable

The main cables on modern suspension bridges are constructed from a single cable, which is spun back and forth around the entire structure. A 18,640-mile strand was used to make the suspension cables for the Forth Road Bridge (1964) in Scotland.

Suspension bridges have existed for thousands of years, but have been perfected in the last two centuries thanks to the advancement of modern materials and construction practices. A suspension bridge relies on the tensile strength of a cable, from which a deck is hung. The earliest bridges to exploit this idea were footbridges suspended from ropes or vines strung across a stream, creating a catenary, or natural, arc. The steel cables used in the construction of modern suspension bridges are anchored to the ground and typically spun in situ over two pylons to create the main span between. The vertical cables connecting the main cables to the deck are called hangers.

Structure

The construction of a suspension bridge begins with the pylons. The main cables are created by spinning a continuous steel wire back and forth over the pylons and through loops at the anchorage. Once complete, these cables are bundled together to create one large cable. The deck beams are then hung on smaller cables. The decks of many modern suspension bridges are box girders. These beams are connected and strengthened to resist vertical loads as well as lateral loads caused by wind. Methods of strengthening vary depending on the type and span of the bridge but can include truss beams, cross bracing, or guy wires.

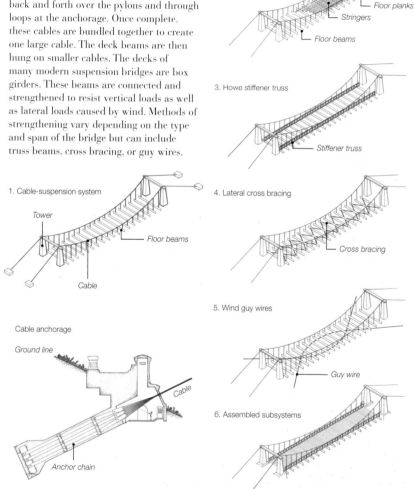

1. Cable-suspension system

Tower

Floor beams

Cable

Cable anchorage

Ground line

Cable

Anchor chain

2. Timber deck

Floor planks

Stringers

Floor beams

3. Howe stiffener truss

Stiffener truss

4. Lateral cross bracing

Cross bracing

5. Wind guy wires

Guy wire

6. Assembled subsystems

Suspension Bridges

Iron and steel

The suspending elements of early suspension bridges were made from iron chains or a series of interconnected flat bars called eyebars. A more effective method of using wire cables was devised in the early 19th century. Cables are stronger and possess more redundancy—a worn wire in a bundle of thousands is not catastrophic, but a fractured eyebar could be. Permanent wire-cable bridges were first built in the early 19th century, but development of steel swiftly produced the first steel-cable suspension bridge: Roebling's Brooklyn Bridge (1883), in New York.

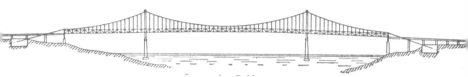

Efficiency

Suspension bridges are the most efficient means of spanning large distances. They can be constructed without obstructing traffic beneath and require no falsework. They also use comparatively little material while being able to carry heavy loads.

Brooklyn Bridge

The steel wires were spun continuously over the pylons, creating four main suspension cables that are anchored into the bedrock of Manhattan and Brooklyn. The Brooklyn Bridge's main span of 1,594 feet made it the longest suspension bridge in the world for two decades.

Failure

Suspension bridges are very effective at withstanding vertical loads, but their slender lines and flexible components expose them to dynamic loads that exceed normal carrying loads, such as crosswinds. These loads, if frequent and not countered in the deck design, can be catastrophic, as with the collapse of the Tacoma Narrows suspension bridge in 1940.

Cable-Stayed Bridges

Versatility

Cable-stayed bridges were first used for situations that were too large for cantilevered bridges and too small to warrant a suspension bridge. However, with improvements in design and construction, cable-stayed bridges are now rivaling suspension bridges, with spans exceeding $\frac{2}{3}$ mile, making them a very versatile bridging solution. The main span in the middle of the cable-stayed Oresund Bridge (2000) that carries railroad and road traffic between Sweden and Denmark is 1,611 foot wide and supported by four independent needle towers.

Cable-stayed bridges appear similar to suspension bridges but are, in fact, more like the cantilever bridge. There are many different types of cable-stayed bridges, but they are characterized by a pylon, or tower, supporting the deck with a series of individual cables in tension. In all cases, the deck is built from the tower in a series of cantilevers. These are arranged on both sides of the tower, which balances the structure and requires no anchorage. Because the deck is being pulled back to the tower and is, therefore, in compression, it has to be a very rigid structure, unlike the suspension bridge, in which the light deck merely hangs from the suspension cables.

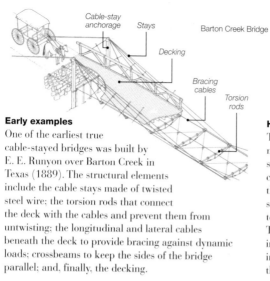

Cable-stay anchorage — Stays

Barton Creek Bridge

Decking

Bracing cables

Torsion rods

Early examples

One of the earliest true cable-stayed bridges was built by E. E. Runyon over Barton Creek in Texas (1889). The structural elements include the cable stays made of twisted steel wire; the torsion rods that connect the deck with the cables and prevent them from untwisting; the longitudinal and lateral cables beneath the deck to provide bracing against dynamic loads; crossbeams to keep the sides of the bridge parallel; and, finally, the decking.

Harp and fan (below)

There are two principal methods of arranging the supporting cables in typical cable-stayed bridges. One is the fan design, in which the supporting cables are gathered toward the top of the tower. The other is the harp design, in which they are arranged in parallel as they rise to the tower.

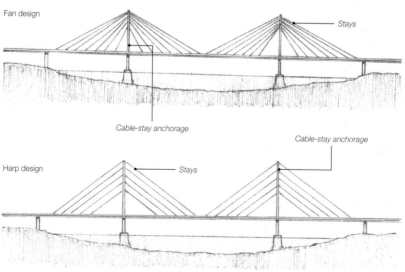

Fan design

Stays

Cable-stay anchorage

Harp design

Stays

Cable-stay anchorage

Cable-Stayed Bridges

Unlimited length

The main spans of modern cable-stayed bridges cannot match those of the longest suspension bridges, but the advantage of cable-stayed bridges is that they can repeat continuously. Because anchor points for cable-stayed bridges lie in the towers, the bridges can be extended by multiplying the towers. The Ting Kau Bridge (1998) in Hong Kong uses a series of towers to traverse a large expanse of water. Extra-long cables, measuring 1,525 feet, increase its longitudinal stability.

Construction (right)

The construction of large, modern cable-stayed bridges begins with the approach roads and tower. Once these elements are in place, the deck is hoisted into position in sections hung from separate cables. When complete, the whole deck is connected and stiffened. Construction is comparatively straightforward and has two main advantages: it requires little or no falsework and there is limited obstruction to shipping beneath.

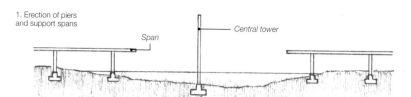

1. Erection of piers
and support spans

Span

Central tower

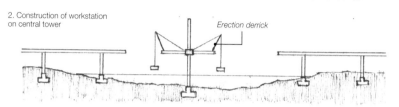

2. Construction of workstation
on central tower

Erection derrick

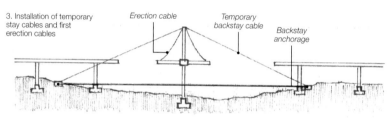

3. Installation of temporary
stay cables and first
erection cables

Erection cable

*Temporary
backstay cable*

*Backstay
anchorage*

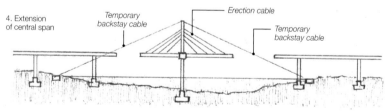

4. Extension
of central span

*Temporary
backstay cable*

Erection cable

*Temporary
backstay cable*

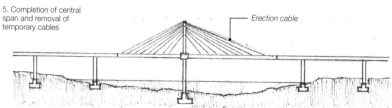

5. Completion of central
span and removal of
temporary cables

Erection cable

Hybrid Bridges

Arch and trusses

The Dom Luís I Bridge (1886) over the Douro River, in Portugal, comprises an arch supporting a railroad deck with a lower road deck hanging beneath. The upper decks are steel-truss beams supported on pylons, with only the central portion being carried by the arch.

Relatively few bridges are true examples of a single bridge type. Most are hybrids of some kind, but some are more explicitly hybrid than others. Hybrid bridges are the most challenging bridges to read because two or more structural types are working in tandem. Certain bridges are designed as hybrids, whereas others become hybrid throughout their working lives, as they are strengthened or altered to meet changing circumstances. The inherently varied character of hybrid bridges means that there is an infinite variety of types. The following are just a selection of examples to illustrate how different bridge types can be used in a single bridge.

Cables and box girders

The decks of cable-stayed bridges are an important structural component of the bridge. They have to withstand the compressive forces imposed by the cables pulling them back toward the towers and the bending forces imposed by the loads pushing them downward. The Erskine Bridge (1971) in Scotland uses cable stays to support the deck, which is constructed from welded, segmental box girders.

Unintentional hybrids

Some bridges are not designed to be hybrids but become so during their lifetime. The Albert Bridge (1873) in London was originally a cable-stayed bridge with thirty-two iron rods supporting the deck. Problems with the bridge led to steel chains, similar to those of a suspension bridge, being added ten years later. In the 1970s, a concrete pier was inserted in the middle of the reinforced span to reduce the load on the towers.

Arches and beams

The engineer Robert Stephenson inserted arches into the wrought-iron ribs to create the individual spans of the High Level Bridge (1849) in Newcastle, England. The distinctive section combining a tied arch with horizontal and vertical ribs, sometimes referred to as a bowstring girder, is extremely rigid and was designed to withstand the loads imposed by this railroad and road bridge.

Hybrid Bridges

Extradosed bridges

By combining the attributes of cable stays and beams, some bridges are able to achieve greater spans and narrower decks than beam bridges and not incur the costs associated with large cable-stayed bridges. Bridges that combine the two are referred to as extradosed and are very distinctive.

The first section of the deck is directly supported on the tower to act as a continuous beam. The cable stays support the next section of the deck, while the central portion of the span is suspended by outer cables. The cable stays of the Ganter Bridge, in Switzerland (1980), are embedded in the prestressed-concrete arm.

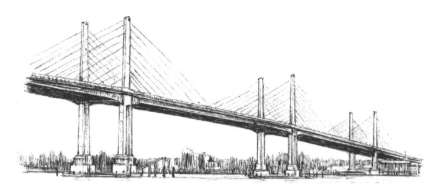

Golden Ears Bridge (above)

The longest extradosed bridge in North America is the Golden Ears Bridge (2009) in Canada, with four towers spanning over 1¼ miles.

Bridge series (below, left)

Some modern bridges are so large that they are a series of bridges rather than a single one. The 26-mile-long Qingdao Haiwan Bridge (2011) in the bay of the Chinese port of Qingdao is the world's longest bridge over water. It comprises cable-stayed sections and both prestressed- and posttensioned-concrete beams.

Cable-stayed section

Prestressed and posttensioned beams

Sunniberg Bridge (right)

Another advantage of the extradosed bridge is that its relatively short spans can create a curved deck. The Sunniberg Bridge (1998) in Switzerland is a reinforced-concrete extradosed bridge with four piers and five spans that curve in an arc through the steep valley.

Introduction

Bridges are designed to satisfy a wide range of different uses from the obvious, such as vehicular, railroad, cycling, and pedestrian traffic, to the more obscure, such as carrying water. Many are even designed to cater to multiple uses.

Most bridges are designed for a specific use (or uses), but the actual use often changes to adapt to the shifting needs of different users. Bridges often acquire a use for which they were never originally intended, which can evolve much later. Ancient bridges that were designed to carry battalions of soldiers or essential water supplies now

bear only the weight of tourists and hikers. Others are victims of their own success and have to be widened or converted from a single deck to a double deck.

Large bridges often have multiple uses. Because they are designed to carry the heaviest loads, they can also accommodate the relatively negligible weight of pedestrians and cyclists. Some bridges have become so iconic that the chance to walk over their structure or to jump from it while attached to a bungee rope becomes a pursuit in its own right.

Embracing the car
Thousands of bridges were built to facilitate the expansion of the road network in the United States. One of the most famous is the reinforced-concrete deck-arched Bixby Creek Bridge (1932) on the West Coast.

BRIDGE USES

Pedestrian

Linking spaces
The Millennium Footbridge (2000) connects St. Paul's Cathedral in the City of London to the South Bank of the Thames River. The high-tech, horizontal suspension bridge has transformed the urban spaces on both sides.

The oldest bridges were designed to be walked over. However, the primitive origins of the footbridge belie the range and depth of its sophistication today. In recent years, the pedestrian bridge has been enjoying a renaissance as urban planning and design begin to retreat from decades of automobile dominance and embrace a more sensitive and humane approach that prioritizes the pedestrian and cyclist. The power of the pedestrian bridge to enhance the urban landscape is being proven all over the world and is resulting in all kinds of innovative and challenging solutions in bridge design.

Romance

Unencumbered by other forms of traffic, pedestrian bridges have the capacity to be places of tranquillity, contemplation, and romance, and sites from which the city or surroundings can be observed from exciting and unobstructed perspectives. Few bridges in the world are more renowned for their romantic possibilities than Venice's Ponte di Rialto (1591), designed by Antonio da Ponte.

Urban regeneration

The Merchant's Bridge (1996) was an integral component of the regeneration of the Castlefield district of Manchester. The design brief demanded a striking and charismatic structure that would transform this formerly neglected part of the city. The solution, designed by Whitby & Bird, was a distinctive 94-ton steel structure characterized by the sickle arch that supports the curved deck.

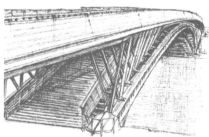

Design innovation

The comparatively small-scale and custom-made nature of pedestrian bridges offers exceptional opportunities for design innovation. The single-span Passerelle Léopold-Sédar-Senghor (1999) in Paris is a steel-arch bridge with a uniquely designed intersecting split-level deck that lets people walk directly over the main arch or on its supported upper deck.

Water

Transport

Aqueducts have helped to boost trade since the 18th century and are still used for transporting raw materials. The world's longest navigable aqueduct is the Magdeburg Water Bridge (2003) in Germany, which is nearly ⅔ mile long.

Bridges that carry water are called aqueducts—a name that reveals its most celebrated usage. By joining together the Latin words *aqua* ("water") and *ductus* ("to lead"), the Romans defined a type of bridge that has continued to be used for thousands of years. The earliest aqueducts were built for irrigation, but as cities grew and trade increased, water was needed for more than just growing crops. The Romans were expert builders of aqueducts and erected them to deliver clean water into cities for drinking and washing. Modern aqueducts are used for transportation and were critical during the Industrial Revolution in linking networks of canals.

Pont du Gard, Nîmes, France (above)

The Romans were the most prolific aqueduct builders, and their engineering skills were not surpassed for more than a thousand years. One of the most famous Roman aqueducts is the Pont du Gard in southern France. Once serving the city of Nîmes, the aqueduct's distinctive profile of three layers of arches stands 157 feet high and 900 feet long.

Aqua Claudia, Rome (below)

No Roman city consumed more water than Rome itself, with its thousands of baths and its position at the heart of the empire. To satisfy this demand, Emperor Caligula ordered the construction of the impressive stone-arched Aqua Claudia in 38 CE; it was completed in 52 CE. This aqueduct supplied water to fourteen districts of Rome.

Industry

Canals were essential to the Industrial Revolution. In designing the canal systems, engineers also had to design the aqueducts to bypass existing obstacles. The Stanley Ferry Aqueduct (1839) in Yorkshire carrying the Aire and Calder Navigation is the longest cast-iron aqueduct in the world.

Vehicular

Banpo Bridge

The Banpo Bridge (1982) across the Han River in Seoul is a major road bridge constructed over another road bridge, the Jamsu Bridge. To enhance this double-deck bridge, the city installed the Moonlight Rainbow Fountain along its length.

Up until the 19th century, most bridges had to carry little more than pedestrians and horse-drawn carts. With the advent of steam power and, later, the combustion engine, vehicular traffic increased exponentially and bridges became larger and stronger in order to bear the weight of modern traffic. Over the last century, the vast majority of bridges were designed to accommodate vehicular transport. Vehicular bridges come in all shapes and sizes. Some small structures provide improved links in established transport networks, while other larger ones create entirely new transport networks, carrying millions of vehicles every year across valleys, bays, or even seas.

Spirit of freedom (right)

No country has embraced the automobile and the spirit of individual freedom that this means of transport represents more passionately than the United States. Bridge building was essential to the expansive road network as it spread westward from the late 19th century. The Bixby Creek Bridge (1932) on the West Coast is one of countless bridges built during the heyday of U.S. highway construction.

Car boom

The 5¼-mile-long General Rafael Urdaneta Bridge (1962) in Venezuela, which combines a cable-stayed central section and prestressed-concrete-beam approaches, is an example of the enormous investments made in highway infrastructure over the last fifty years.

Rokko Island Bridge

The 712-foot-long, steel, tied-arched Rokko Island Bridge (1993) in Kobe, Japan, forms part of a busy expressway and is another example of a double-deck road bridge designed to carry large volumes of urban traffic.

Rail

Glenfinnan Viaduct
The Glenfinnan Viaduct (1901) in Scotland was the first in the world to be built entirely from concrete. The structure, with its twenty-one arches, was built by Sir Robert McAlpine using nonreinforced concrete.

With the advent of the railroads in the 18th century, new methods of construction had to be devised that could carry these unprecedented loads. The first railroads did not carry passengers but heavier cargo, such as coal and other raw materials that fueled the Industrial Revolution. The oldest surviving rail bridge is Causey Arch (1727) in County Durham, a stone arch erected by the mason Ralph Wood. Since these early beginnings, rail bridges have been built worldwide, using every method of construction, to carry heavier loads and cross ever wider distances.

Del Garda Viaduct

Viaducts are among the most celebrated and beautiful types of structures designed to carry trains. The pointed-arch stone viaduct on the Del Garda Railroad in Italy was built in 1852 but destroyed by air raids during World War II.

Garabit Viaduct

The Garabit Viaduct (1884) in France was built by one of the most famous engineers of the 19th century, Gustave Eiffel. Dominated by the 540-foot steel-truss arch that leaps across the Truyère River, the entire length of the railroad bridge is more than 1,640 feet. The use of trusses was deliberate to minimize wind resistance.

Bridge over the Kwai

Linking Thailand and Burma, the bridge over the Kwai Yai River was one of the greatest feats of engineering undertaken by the Allied prisoners who built the railroad line during World War II. The bridge was immortalized in Pierre Boulle's novel *The Bridge over the River Kwai*, which was made into a movie in 1957.

Military

The Bailey bridge
The Bailey bridge was designed in Britain during World War II. The success of the steel-truss design was due to its transportability and ease of construction by hand and often under fire.

Bridges designed and constructed specifically for military purposes are a unique breed. Unlike other types of bridges, they are temporary because it is essential for them to be easily and quickly constructed and deconstructed, as well as to be transportable. Military bridges have been used for thousands of years and are often critical to strategic success. The earliest types of military bridges were of the pontoon variety, in which buoyant sections are bound together to form the deck. Today, with the weight and size of modern military hardware, military bridges are often a source of extraordinary innovation.

Modular design

The modular design of 10-foot
sections gains its strength from
the lattice truss that forms the
side of the bridge, each of which
weighs 573 pounds and requires just six
men to construct it. A testament to its
enduring design, the Bailey bridge is still
used today in peacetime and sometimes
even as a permanent structure.

Specialty bridges (above)

Modern armies are equipped with a wide
range of tools that can create or perform the
function of bridges. This 11½-foot-wide
and 66-foot-long bridge is made from a
lightweight alloy and can be folded in half
and carried on a converted tank.

Pontoon (below and right)

The pontoon is the oldest and the classic
example of military bridge design. A
series of floating sections made of boats,
inflatables, or natural materials, such as
bamboo, supports a lightweight deck.
Some form of cross bracing is usually
required to provide enough structural
rigidity to prevent the bridge from being
washed away by fast-flowing rivers.

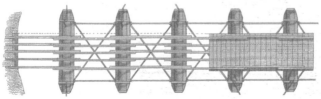

ENGINEERS Introduction

The majority of bridges are rarely noticed, let alone the talents of those who designed them. The names of the countless men and women who have designed and overseen the construction of bridges are invariably forgotten, but occasionally a bridge will make (or even break) a reputation and career. Through the centuries, great bridge designers have helped to transform both their trade and their society, advancing commerce and communications regionally or improving the urban realm locally.

Some of the most prolific and talented bridge designers have exploited the unprecedented developments in materials and construction technologies over the last two

centuries. Conventionally, most of these often-celebrated figures have been engineers, but their ranks also include architects, artists, and even sculptors. Common to each of these individuals is a talent for designing effective and efficient bridges that do not merely provide a connection between two points but inspire and thrill the user or onlooker in the process.

The following is a small selection of the hundreds of influential bridge designers who have helped not only to transform and improve the world we live in, but who have also inspired generations of fellow professionals and members of the public through the beauty of their work.

Victorian engineer
Few engineers are more intimately associated with the advancement of bridge design than Isambard Kingdom Brunel. Below is his Royal Albert Bridge (1854–59), constructed of wrought iron, using a pair of lenticular trusses supported on masonry piers.

Isambard Kingdom Brunel

No other individual in Britain made such a marked contribution to civil engineering as Isambard Kingdom Brunel (1806–59). His prodigious list of works includes tunnels, bridges, boats, buildings, and railroads. Brunel gained his early engineering experience working on the construction of the Thames Tunnel for his father, the engineer Marc Isambard Brunel. Brunel's diverse career was dominated by his work on the Great Western Railway between London and Bristol. He also designed transatlantic ships, including the first engine-powered and propeller-driven metal ship, the SS *Great Britain* (1843).

Chepstow
Brunel's economical design for the 302-foot-span railroad bridge over the Wye River at Chepstow (1852) uses a bowed tubular girder to support a rigid iron deck using iron chains. His novel design was a prototype for the larger and more accomplished Royal Albert Bridge.

Wharncliffe Viaduct
One of Brunel's first designs for bridges along the Great Western Railway was the 885-foot-long Wharncliffe Viaduct (1836–37) west of London, between Hanwell and Southall. The brick bridge comprises eight semielliptical arches supported on hollow, tapered piers.

Hungerford Bridge (above)

The Hungerford Bridge (1845) was an iron suspension footbridge across London's Thames River. The bridge was purchased by the South Eastern Railway Company in 1859, who dismantled it when building Charing Cross Station. The new railroad bridge used Brunel's brick piers, and the bridge's iron chains were reused on Brunel's famous Clifton Suspension Bridge (see pages 198–199).

Royal Albert Bridge (below)

The Royal Albert Bridge (1854–59) is a railroad bridge that links the counties of Devon and Cornwall across the Tamar River. The wrought-iron structure comprises two lenticular trusses (see page 47), each 453 feet long. These are connected to approaches made of a continuous iron-plate beam supported on stone piers, giving the bridge a total span of 2,185 feet.

John A. Roebling

Brooklyn Bridge
The largest of
Roebling's bridges
is the 1,594-foot-
long Brooklyn
Bridge (1883).
When completed,
it was the longest
suspension bridge
in the world. In
1869, Roebling
died from tetanus
poisoning after a
ferry crushed his
foot while doing
survey work. His
son, Washington
Roebling, took over
the job, but he was
later paralyzed after
suffering from
decompression
sickness while
working on the
caissons that kept
the foundations free
of water during
their construction.
John Roebling's
daughter-in-law,
Emily Warren
Roebling, then
continued
the project.

John A. Roebling (1806–69) spent the first twenty-five years of his life in Germany, where he studied and trained as an engineer. In 1831, he went to the United States, where after a few years working as a farmer, he returned to engineering. He specialized in suspension bridges, having studied them in Germany, and manufactured wire cables on his Pennsylvania farm. His first suspension bridges were aqueducts, after which he went on to design much larger bridges, culminating in his most famous work, the Brooklyn Bridge across New York's East River.

Niagara Falls Suspension Bridge

In 1851, Roebling started work on his first major suspension bridge, which crossed the Niagara River as part of the new railroad between New York and Canada. The 823-foot-span bridge was groundbreaking in its use of metal wire cables to support the double deck, which carried trains above and vehicles below.

Overhead view of Delaware Aqueduct

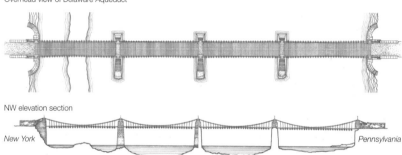

NW elevation section

New York   Pennsylvania

Delaware Aqueduct (above)

Roebling honed his understanding of suspension bridges on four aqueducts he designed for the Delaware and Hudson Canal from 1848. This particular bridge has four separate spans supported on three piers.

Roebling Suspension Bridge (below)

Roebling's 1,056-foot-long Cincinnati–Covington Bridge (later renamed the John A. Roebling Suspension Bridge) over the Ohio River was the longest suspension bridge in the world when it opened in 1866, and was a precursor to his Brooklyn Bridge.

Robert Maillart

Salginatobel Bridge
Salginatobel Bridge
(1930) shows
Maillart's innovative
use of concrete.
His intuition for
and understanding
of concrete's
behavior made his
structures not only
beautiful but cheap
and quick to build.

The first person to fully embrace concrete's aesthetic and structural potential in bridge design was the Swiss engineer Robert Maillart (1872–1940). Although Maillart trained at Zurich's Federal Institute of Technology, he was more an intuitive engineer than a methodical one. His ability to manipulate concrete, which at the time was a new and little-understood material, was revolutionary. Schwandbach Bridge and Salginatobel Bridge (see pages 128–129) are his most renowned and radical works, influencing future generations of bridge designers.

Stauffacher Bridge

One of Maillart's earliest bridges was the Stauffacher Bridge (1899) in Zurich, Switzerland. The bridge appears to be constructed out of stone but is, in fact, fabricated from reinforced concrete using a hollow box above a three-hinged arch. The deception illustrates Maillart's early work before he fully grasped the structural freedom and simplicity of reinforced concrete.

Bohlbach Bridge

In the Bohlbach Bridge (1932), Maillart developed the deck-stiffened-arch technique. The reinforced-concrete deck sits on a slender arch supported by slim piers the same width as the arch and deck. The road's banked curve, which counters the centrifugal forces created by traffic, is accommodated by varying the width of the arch.

Schwandbach Bridge

The zenith of Maillart's use of the deck-stiffened arch was at Schwandbach (1933), where the polygonal arch beneath the stiffened deck is approximately 8 inches thick. As with Bohlbach, the road passing over the bridge is curved, causing the width of the arch to vary from 13 feet in the center of the arch to 20 feet at the abutments.

Santiago Calatrava

The work of Santiago Calatrava (b. 1951) seamlessly traverses the disciplines of engineering, architecture, and sculpture. His famous structures, which include bridges, skyscrapers, and railroad stations, have transformed and reinvigorated public appreciation of engineering in the 20th and 21st centuries. His bridges are celebrated for their distinction and sculptural quality.

Jerusalem Chords
The cantilever spar cable-stayed bridge, which Calatrava first used in his design for Seville's Puente del Alamillo (1992), has become one of his trademarks. The technique is a variant of the standard cable-stayed bridge in which the mast is angled to counter the load of the bridge, requiring fewer cable stays. Calatrava used the cantilever-spar method for the Jerusalem Chords Bridge (2005–08) (carrying trams, vehicles, and pedestrians) because of the bold visual statement it makes.

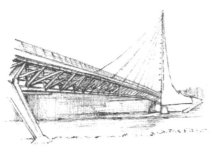

Sundial Bridge

In his design for the bicycle and pedestrian footbridge over the Sacramento River at Redding, California, Calatrava transforms the 217-foot mast into a sundial's gnomon. The clock face forms a plaza on the north side of the bridge. The fourteen cable stays of the Sundial Bridge (2004) use more than ⅔ mile of cable to support the 700-foot span.

Ponte della Costituzione

The Ponte della Costituzione (2007–08) is an arched truss comprising five arched members: one in the center, two underneath the bridge, and two on each side. With a 262-foot span, these members are connected with steel tubes and plates that form a sculptural ribbed structure supporting the stone and glass steps of the bridge. (See also pages 154–155.)

Samuel Beckett Bridge

Calatrava modified the cantilever spar cable-stayed bridge in his design for Dublin's Samuel Beckett Bridge (2009) by using an arced tubular-steel mast supported by cables on both sides.

The road and pedestrian bridge swings 90 degrees to let boats on Dublin's River Liffey pass. Although the bridge was designed to be a modern landmark, its form also evokes the traditional Irish harp.

Gustave Eiffel

After training at the prestigious École Centrale des Arts et Manufactures in Paris, Gustave Eiffel (1832–1923) went on to become France's most famous engineer. He designed many groundbreaking structures, the most famous of which is the tower he designed and built for the Paris Exposition Universelle in 1889, which bears his name. Intended to be a temporary structure, the Eiffel Tower captured the hearts and minds of Parisians and is the city's most famous icon. Eiffel also designed the armature inside the Statue of Liberty (1886), which was a gift from France to the United States.

Garabit Viaduct

The Garabit Viaduct (1884) across the Truyère Valley in southern France advances Eiffel's earlier designs, including the Dom Luís I Bridge (1875) in Portugal (see page 64). The 1,854-foot-long steel bridge is dominated by the 540-foot-span crescent arch, on which the 407-foot-high truss deck sits. It was the highest bridge in the world at the time.

Maria Pia (left)

Another of Eiffel's early designs is the two-hinged, crescent-arched Maria Pia Bridge (1877) in Portugal. As with the Dom Luís I Bridge, the wrought-iron truss deck of the Maria Pia is incorporated into the crown of the arch, whose 525-foot span was then the longest in the world. Eiffel was a pioneer in the use of trusses, which he exploited because they reduced wind resistance compared with solid beams.

Zrenjanin (right)

The Great Bridge (1904) in the city of Zrenjanin in Serbia uses a small steel truss to cross the Begej River. Eiffel designed the bridge so that it could be raised and lowered to let river traffic pass underneath. The bridge was replaced in 1969, but plans are underway to rebuild it.

Cubzac-les-Ponts

The Cubzac-les-Ponts (1883) in southern France comprises a series of box trusses supported by seven pairs of cross-braced, round wrought-iron piers on stone footings. The approaches are built on a masonry viaduct with longitudinal pointed-arch vaulting. Eiffel's road bridge is 1,800 feet long and weighs more than 3,300 tons.

Benjamin Baker

ENGINEERS

Partnership
After working for Wilson, Baker joined the engineer John Fowler, and they went into partnership in 1875. Together they worked on London's Metropolitan Railway, the world's first urban rail network, and the Forth Rail Bridge (1882–90), in Scotland.

As a young engineer, Benjamin Baker (1840–1907) was apprenticed to the renowned Neath Abbey Iron Works in Wales, after which he became a prolific writer and learned teacher of engineering and one of the greatest engineers of the Victorian age. As assistant to Mr. W. Wilson from 1860, Baker worked on the construction of London's Victoria Station and Victoria Bridge before working on the first underground railroads. His contributions to engineering span the globe and include tunneling for the London Underground, the Hudson River Tunnel in New York, and the building of Egypt's Aswan Dam. The one design for which he will always be remembered is the Forth Rail Bridge.

Troubled start

The Forth Bridge Company was formed in 1873 and accepted a suspension-bridge design by Sir Thomas Bouch, engineer of the Tay Bridge. Unfortunately for Bouch, the Tay Bridge collapsed in 1877, along with public confidence. His Forth suspension bridge was scrapped and Fowler and Baker's design for a steel-cantilever and central-girder bridge comprising three massive double cantilevers was accepted.

Construction

Fabricated from more than 71,000 tons of steel (the first time that steel was used so predominantly in a bridge), the bridge's double cantilevers sit on granite piers and iron caissons. They are linked by short suspended trusses creating two main spans more than 1,640 feet long—the longest in the world until 1917.

Introduction

A beam bridge is a structure comprising a horizontal element supported at both ends by piers. The weight of objects moving over the beam (live loads) combined with the weight of the beam (dead load) create vertical forces that are transferred through the structure and supported by the piers. The ability of the beam to withstand these forces may be judged intuitively by the span-to-depth ratio, which is defined as the relationship between the depth of its section and the distance between the piers. If the span is too wide and the section too thin, the beam may fail.

There are two types of beam bridges: a simply supported beam and a continuous beam. The former is the basic type, in which a single-beam structure is supported at each end by piers. The latter comprises a beam structure that is continuous over the supports. The thickness of a continuous beam can be reduced in section because it works more efficiently by developing hogging moments in its cross section over the piers as well as sagging moments over its span. Beam bridges are among the longest in the world.

Alter ego

The Rio–Niterói Bridge, in Brazil, is officially known as the President Costa e Silva Bridge. At more than 8 miles long, it was the second longest bridge in the world when completed in 1974.

Anping Bridge

Anping Bridge is an old beam bridge built in the 1140s in Fujian Province, in southeast China. Extending a distance of 7,400 feet, the 10-to-13-foot-wide bridge was for many centuries the longest beam bridge in the world, and it was the longest bridge in China until 1905. Each section of the bridge is made up of six stone beams that are laid next to one another and supported at each end by a stone pier. Three pavilions and large carved figures once stood along the bridge.

Endurance
Despite its age, the Anping Bridge is still in use and well maintained by local craftsmen. It is open to pedestrians and cyclists but not to vehicular traffic.

Beams

The stones that make up the bridge's sections are rectangular beams with a thickness in section large enough to support their own weight of over 27 tons. The weight of each beam indicates the structure's inefficiency, where much of the structural capacity is used in supporting its own weight instead of the live loads.

Balustrade (right)

A stone balustrade lines the entire length of the bridge. Each deck section has three sections of balustrade. Note how the large load-bearing granite posts sit in the space between the beams on each pier, with two shorter intermediary posts between them.

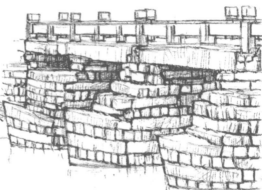

Piers

The bridge has a total of 331 piers along its entire length. These can be arranged into approximately three different types. Some are rectangular in plan, some are slightly pointed to make them more streamlined, and others, resembling the shape of a boat, are completely streamlined at both ends to reduce resistance to the river. All are built from layers of stone beams, perpendicularly arranged to bond the structure together.

Britannia Bridge

Former beam
This image of
Britannia Bridge
was taken before
the bridge was
damaged by a fire
inside the tubular
beams in 1970. It
was strengthened
using truss arches,
which replaced the
tubular beams.

The Britannia Bridge (1850) crosses the Menai Straits between mainland Wales and the island of Anglesey. The original bridge was a revolutionary piece of engineering by Robert Stephenson (1803–59). His radical solution developed a previous design of a road bridge that had used a wrought-iron trough. By enclosing the trough, Stephenson was able to create a strong tubular girder through which the railroad track could pass. The series of girders were supported on masonry piers, but in 1970 a fire in the bridge weakened the structure and it was rebuilt using truss arches on the same piers.

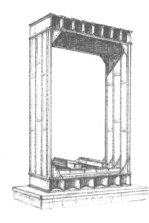

Tubular girder

The design and construction of the tubular girder was essential to the bridge's strength. Each section was manufactured from wrought-iron sheets riveted together with a smaller boxed section along the top and bottom to provide additional strength and stiffness. The trains ran inside the beam.

Lifting the beams

It was a condition that the bridge construction did not affect the navigation channel beneath. The decision to use beams allowed for the piers to be built independently, then each section was lifted into place using hydraulic jacks, a process that took a total of seventeen days.

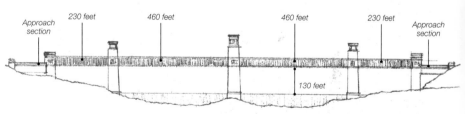

Approach section | 230 feet | 460 feet | 460 feet | 230 feet | Approach section

130 feet

Bridge length

The beams that create each of the bridge's two 460-foot main spans were considerably longer than any beam that had been built previously. Each span section weighed over 1,600 tons and was supported by tall masonry piers that gave the bridge a clearance of 130 feet. Two shorter 230-foot-span sections connect the bridge from the island of Anglesey to mainland Wales. The total length of the bridge, including the shorter approach sections, is 1,500 feet.

Tay Rail Bridge

Piers

The arched approaches of the Tay Rail Bridge stand on solid brick piers, while the piers supporting the truss beams are concrete arches standing on two legs that are tied toward the base by a horizontal beam.

The original Tay Bridge, completed in 1878, used a system of simply supported truss beams fabricated from a latticework of cast and wrought iron. The following year the bridge collapsed during a major storm, causing Britain's biggest railroad disaster of the Victorian era, in which all 75 passengers and crew were killed. The new bridge (1887) was constructed with two different types of simply supported beams: one was rectangular with the railroad track above, and the other arched with the track passing between the arches.

Assorted trusses

Unlike the approaches, which are built from lattice trusses underneath the deck, the central section of the bridge is formed of arched-bowstring trusses set above deck and simply supported on piers at each end. This arrangement increases the clearance by extending the distance between piers and the height of the deck above water to accommodate the navigational channel.

Truss beams (below)

On the approaches to the central section of the bridge, the deck is supported on four parallel steel-lattice trusses. The trusses are stabilized by cross bracing and their ends sit on cast-iron bulkheads that transfer the load down to two piers.

Joint

Individual members are riveted together to form the latticework. Small vertical members attached to the intersection of the diagonal further strengthen the truss. Expansion joints between the individual beams, where they sit on the piers, indicate that this is a simply supported structure instead of a continuous-beam one.

Lake Pontchartrain Causeway

The economical benefit of using simply supported beams for very long bridges is nowhere better illustrated than on the Lake Pontchartrain Causeway in Louisiana. The causeway comprises two 24-mile-long parallel road bridges (northbound and southbound), one built in 1956 and the other in 1969. Except for the central bascule section that allows for boats to pass through, the entire length of both bridges is constructed using prestressed reinforced-concrete beams simply supported on concrete piers. Each beam was prefabricated and taken to the site on a barge, from where it was lifted onto the piers by floating cranes.

Prefabrication

The use of prefabricated reinforced-concrete sections in the construction of a bridge this long significantly reduces construction times and cost because each section can be fabricated off-site and then secured into place with relative ease and economy.

Expansion joints

With so many beams laid end to end, it is necessary for there to be a small gap between them to let the structure expand and contract. One end of each beam is held securely on a perpendicular crossbeam supported on piers with the other end free to expand in the direction of the span.

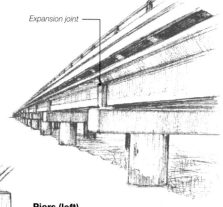

Expansion joint

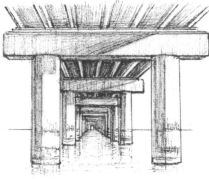

Piers (left)

Over 9,500 reinforced concrete piers were used in both bridges. These piers are arranged in pairs, except in the approaches to the central section of the bridge, where the taller piers are arranged in threes, increasing the strength of the structure. Each section of the bridge comprises a pair of piers supporting a group of seven beams.

Gateways

Long bridges over wide water channels seldom provide navigational clearance across the full width because this is not necessary for shipping nor is it economical. Many long bridges over water are, therefore, low lying with raised or moving sections to provide a gateway through which large vessels can pass.

For this purpose, the center of the Lake Pontchartrain Causeway has a raised bascule section (see page 49).

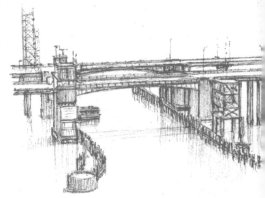

Chesapeake Bay Bridge

Hybrid structure
The Chesapeake Bay Bridge uses three types of bridge systems: reinforced-concrete beams (foreground), steel-truss cantilever (middle), and suspension (background).

The 17-mile crossing of the Chesapeake Bay in Virginia was one of the most complex bridge-tunnel systems in the world when completed in 1964. The main section of the bridge is a 3,200-foot suspension bridge, with steel cantilevered trusses leading up to it. The system also has two tunnel sections that are 1 mile long. The majority of the bridge comprises steel truss beams and concrete beams on trestle piers. A second bridge was built parallel to the Chesapeake Bay Bridge-Tunnel, which opened in 1999.

Cantilever beam

The approach sections to the suspended portion of the bridge feature steel cantilevered trusses. These members act like beams but have a cantilevered form over their support piers, indicated by the profile. The purpose of these members is to raise the deck from the low-lying trestles just 39 feet above sea level to the suspended section, which has a clearance of 184 feet.

Precast beams (below)

Precast reinforced-concrete beams create the road deck along the trestle sections of the bridge. Each deck section comprises three beams supported on concrete trestle piers at both ends, which transfer the loads to the ground. Precasting allowed for the beams to be manufactured onshore and transported out to sea, where they were lifted into place by a floating crane.

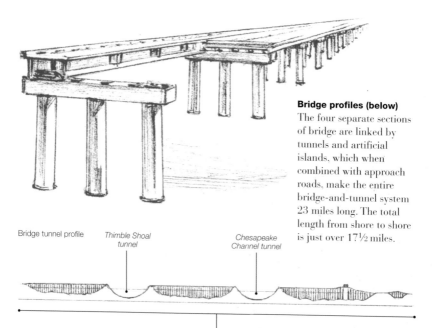

Bridge profiles (below)

The four separate sections of bridge are linked by tunnels and artificial islands, which when combined with approach roads, make the entire bridge-and-tunnel system 23 miles long. The total length from shore to shore is just over 17½ miles.

Bridge tunnel profile

Thimble Shoal tunnel

Chesapeake Channel tunnel

17½ miles shore to shore

Metlac Bridge

Extended deck

The road deck of the Metlac Bridge extends beyond the edge of the beam and is supported on cantilevered brackets projecting from the side of the beam. The projection of the railroad deck is less pronounced.

The railroad bridge and road bridge that cross the Metlac River in Mexico demonstrate two different types of continuous beam. The 420-foot-high steel road bridge was the first to be constructed, and it opened in 1972. It was built using a continuous steel beam carrying four lanes of traffic supported on a series of concrete piers. The longest span between its two piers is 400 feet. The 430-foot-high, two-track railroad bridge opened in 1984. The deck is a concrete box section. The longest span between its five piers is 295 feet.

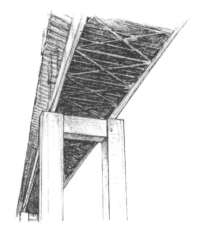

Beam and pier (left)

The narrow beam of the railroad bridge is supported on single piers. The piers of the wider road bridge are arranged in pairs linked at the top by crossbeams. The load is transferred directly onto each pier through the steel girders that form the sides of the beam. The girder is strengthened at its base by a slight thickening of the bottom section of the I beam as it meets the pier.

Continuous beam (right)

The rail bridge's deck is a continuous concrete beam. The piers taper toward their base and were built first, then each beam was constructed as balanced cantilevers before being connected in the center of the spans.

Steel girder (left)

The depth of the steel girder is dictated by the length of the span between the support piers. This continuous steel girder is fabricated from individual sections, which are then welded together and stiffened on the inside by the lateral cross bracing. Lateral and diagonal cross bracing stabilizes the steel girders.

Introduction

Vertical forces acting on an arch are transferred through its curve into the abutments at each side. The strength of the bridge relies on the shape of the arch and the ability of the abutments to resist the vertical and horizontal thrust forces. Whether pointed, round, or segmental, all shapes of conventional arch perform the basic function of transferring vertical forces through a curved or stepped profile to the abutments. Different types of arches include deck-arch bridges, which carry a horizontal deck directly over the arch; through-arch bridges, which support the

horizontal deck between the abutments by suspending it from the arch; and half-through arches, which use a similar principle, but in this instance the deck pierces the arch above the abutments. Tied arches do not rely on the abutments at all to resist the horizontal forces in the arch. Instead, these forces are resisted as tensile forces in the deck, which is tied to the arch, in the same way that a bow maintains its shape by being tied at either end to the bowstring. Unlike other types of arch, the tied arch maintains its structure independently of the abutments.

Pont du Gard
Note how the supporting piers in the Pont du Gard, France, are aligned vertically from top to base. To maintain these proportions, the widest of the larger arches has four small arches above it where the others have three.

Alcántara Bridge

Voussoirs

Note how the voussoirs in each of Alcántara Bridge's round arches are arranged radially, contrasting with the horizontal layers of stone used in the rest of the structure.

Inscribed on the central archway of the bridge across the Tagus River in Spain, built under the orders of the Roman emperor Trajan, is (in Latin) the quotation "I have built a bridge that will last forever." Despite some incidents of destruction in recent centuries, Trajan's ambitious claim has proven correct. The stone structure completed in 106 CE and comprising six round arches (three of which span the river), has stood for nearly two thousand years.

Streamlining

The stone piers supporting the arches midstream are necessarily substantial. To reduce resistance when the river is flooded, the upstream face of each pier is pointed to make it streamlined. Note how the side of the pier facing downstream has not been streamlined. Rounded buttresses extend from each of these pointed profiles to the full height of the bridge to provide lateral stiffness to the sides of the bridge.

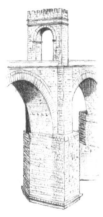

Triumphal arch

A small triumphal arch stands in the center of the bridge. The sides of this smaller lateral arch stand on the buttressing that provides lateral strength to the structure and transfers forces through the piers to the stone foundations.

Symmetry

The six round arches have been built symmetrically. The outermost arches are the smallest, followed by two medium arches, and, in the middle of the bridge, the two largest arches, which span the river.

Zhaozhou Bridge

Decoration

Traditionally, Chinese bridges were often adorned with decorative motifs or small pavilions. Here, ornate stone balustrades are topped with carved figures on the posts.

Built between 595 and 605 CE in China's Hebei Province, the Zhaozhou Bridge is the oldest stone open-spandrel bridge in the world. The 30-foot-wide and 165-foot-long bridge is a segmental arch with a span of 121 feet. The bridge's lightweight design is remarkable for its age. The shallow arc and the open spandrels decrease the bridge's dead load by minimizing the stonework, which consequently reduces the size of the abutments and the force on the arch. The combined use of limestone and iron dovetail joints is also very innovative.

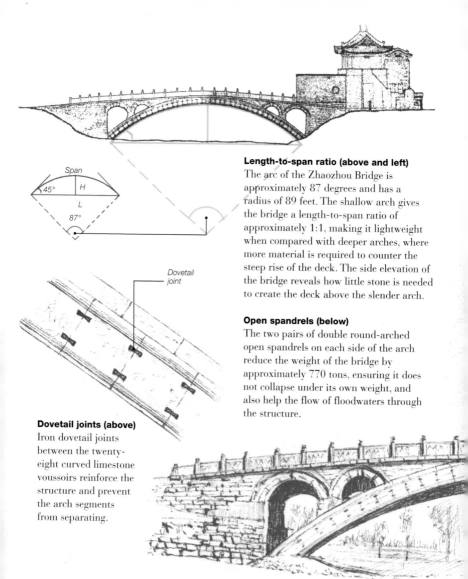

Length-to-span ratio (above and left)

The arc of the Zhaozhou Bridge is approximately 87 degrees and has a radius of 89 feet. The shallow arch gives the bridge a length-to-span ratio of approximately 1:1, making it lightweight when compared with deeper arches, where more material is required to counter the steep rise of the deck. The side elevation of the bridge reveals how little stone is needed to create the deck above the slender arch.

Open spandrels (below)

The two pairs of double round-arched open spandrels on each side of the arch reduce the weight of the bridge by approximately 770 tons, ensuring it does not collapse under its own weight, and also help the flow of floodwaters through the structure.

Dovetail joints (above)

Iron dovetail joints between the twenty-eight curved limestone voussoirs reinforce the structure and prevent the arch segments from separating.

Span

45° *H*

L

87°

Dovetail joint

Lugou Bridge

Dry river
Until relatively recently, the Yongding River passed underneath the Lugou Bridge, but water diversion in the Beijing region has meant that the riverbed is now invariably dry.

Originally built in the 12th century, the Lugou Bridge in China was reconstructed in 1698. It is known outside China as the Marco Polo Bridge because the famed Venetian traveler is said to have praised it highly. The 873-foot-long and 30-foot-wide structure comprises eleven segmental granite arches on ten stone piers between abutments at the bank. Carved lions sit on the pillars of the stone balustrade that lines the bridge's deck. It was on this bridge in July 1937 that the "Incident" between the Chinese and Japanese armies took place, marking the start of the Second Sino–Japanese War.

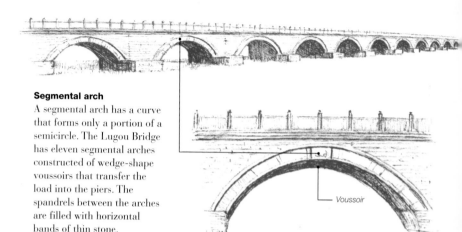

Segmental arch

A segmental arch has a curve that forms only a portion of a semicircle. The Lugou Bridge has eleven segmental arches constructed of wedge-shape voussoirs that transfer the load into the piers. The spandrels between the arches are filled with horizontal bands of thin stone.

Voussoir

Streamlined piers

The piers on the upstream side of the bridge have been designed with pointed heads to protect them from the force of the river that once flowed beneath the bridge. Iron bars have also been added to the heads to help protect the stonework from the water and ice.

Ornament

There are so many lions carved into the 280 balustrade posts that the exact quantity is not known, although it is somewhere between 482 and 496. Each lion is different, and its posture denotes its age of production.

Pont del Diable

Reconstruction
Built on Roman foundations and having survived seven centuries, the medieval Pont del Diable spanning the Llobregat River was destroyed in 1939 during the Spanish Civil War. It was rebuilt in 1965 in a manner faithful to the previous design.

A bridge has crossed this section of the Llobregat River in Catalonia, Spain, since Roman times. Medieval stonemasons used the Roman foundations to build a new bridge in 1283. The bridge is distinctive for having two different types of arches. The smallest is a narrow round-arch opening in the spandrel at the eastern end of the bridge. The medium and largest arches are pointed arches. The bridge's shape reflects the loads on the structure, which are smallest at the center of the main span, where it is thinnest, and increase toward the outer spans, which are supported by the abutments.

Triumphal arch (left)

A Roman triumphal arch stands at the eastern end of the bridge. It was common for such features to appear on important civic structures throughout the Roman Empire. Other Roman remains were used in the foundations on which the medieval bridge was built.

Chapel (right)

A chapel has been built on the apex of the pointed arch. The round-arch stone structure is a distinguishing feature of the bridge and mirrors the triumphal arch on the eastern abutment. The shallow depth between the arch and the chapel's pitched roof helps to reduce the load over the main arch.

Pointed arch (left)

The main arch has a span of over 121 feet, and its pointed form reflects the influence of Gothic architecture at the time it was constructed. The smaller pointed arch has a shorter 62-foot span.

Ponte Vecchio

Living bridges

The Ponte Vecchio has hosted all kinds of stores, from florists to tanners. In 1593, only jewelers were permitted on the bridge. Today, people live in apartments above the stores, which overhang the sides supported by struts.

An advantage of segmental arches or flat arches over round arches is their more efficient span-to-rise ratio. By using segmental arches instead of the round arches favored previously by the Romans, the 14th-century builder Taddeo Gaddi was able to build the Ponte Vecchio on just two piers across the Arno River in Florence, Italy, in 1345. Having fewer piers was more efficient, because there was less obstruction for boats and floodwaters. The width of the bridge's deck is 105 feet, and it has always supported stores or market stalls. From the late 16th century to the early 19th century, other structures have continued to be added to the bridge.

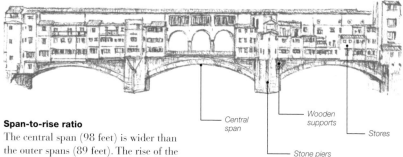

Central span

Wooden supports

Stores

Stone piers

Span-to-rise ratio

The central span (98 feet) is wider than the outer spans (89 feet). The rise of the arches is between $11\frac{1}{2}$ feet and $14\frac{1}{2}$ feet, giving the arches a span-to-rise ratio of approximately 5:1, reducing the number of piers required to bridge the river. This reduces the bridge's resistance to floodwater and minimizes obstruction to river navigation.

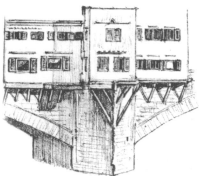

Profiled pier

The face of each pier is profiled to reduce resistance against the water flow, particularly when the river is flooded. The bridge was nearly destroyed in a flood in 1966; it swept through many of the buildings but was not high enough to reach the corridor marked by a series of small windows that passes over the top. The three large windows in the center were created by Benito Mussolini in 1939.

Pulteney Bridge

Picturesque
Pulteney Bridge is one of just four bridges to have buildings along both sides of its entire length. The south facade forms a picturesque view of the bridge with the three-tiered waterfall in the foreground.

Pulteney Bridge (1773) over the Avon River in the center of Bath was Britain's answer to Italy's covered Renaissance bridges and, in particular, Andrea Palladio's unrealized design for the Rialto Bridge over Venice's Grand Canal. Inspired by the reinvention of Classical architecture, the Scottish architect Robert Adam designed the arched stone bridge to connect the city of Bath with the undeveloped land across the river that existed at the time. The bridge crosses the river on two streamlined piers supporting three round arches.

Stores (above)

The inside facade of the bridge facing the road is lined with a row of small stores. The architecture of this facade has been much altered since the bridge was originally built.

Evolution (above)

Adam's original design of Pulteney Bridge lasted just nineteen years. The facade was altered in 1792 to accommodate enlarged stores, and in 1799 severe floods damaged the bridge, requiring it to be substantially rebuilt. Subsequent incremental developments have seen the stores extend out from the north facade on cantilevered supports. The result is a disorderly appearance that contrasts with the original symmetrical design that can still be seen on the south face.

Urban planning

The construction of Pulteney Bridge was an important element of Bath's substantial urban development in the 18th century. Much of the new city was designed along Classical lines by the father and son architects both named John Wood. Adam continued this tradition in his original design.

Maidenhead Railway Bridge

Engineering marvel
The width of the two main spans of Brunel's bridge was an engineering marvel for its time. Three of the thirteen arches of the older Maidenhead Road Bridge (1777) can be seen in the background.

In his role as chief engineer to Britain's Great Western Railway, Isambard Kingdom Brunel designed the world's longest and flattest arch bridge in Maidenhead, just west of London, in 1838. The Great Western Railway's board were not convinced that such a shallow-arch structure could remain standing, let alone withstand the heavy load of the railroad. Consequently, they insisted the wooden formwork supporting the two arches during construction be left in place. Brunel is said to have lowered it slightly so that it had no structural function and soon afterward it was washed away in a flood. The distinctive brick bridge has remained standing ever since.

Arch series

The entire bridge comprises eight arches. The two shallow arches forming the main spans over the river are flanked by three smaller round arches on land on each bank.

Arch width (right)

The original design accommodated two tracks of Brunel's wide-gauge railroad, which was later abandoned in Britain for the narrower standard gauge. The brick bridge was later widened to carry four tracks of the narrower gauge.

Flat arch

The combined length and flatness of the arch allowed for the bridge to cross the Thames River in just two spans of approximately 128 feet each, while preventing a steep gradient in the railroad that would have been created by a taller arch. The total rise measures approximately 23 feet.

Plougastel Bridge

Replacement
A new cable-stayed bridge (visible in the background of this photo) has been built near the old Plougastel Bridge, which is now only used by pedestrians, cyclists, and tractors.

Reinforced concrete was used to build the three large segmental arches of the Plougastel Bridge (1930) crossing the Elorn River in northwestern France. Designed by the civil engineer Eugène Freyssinet, the double-deck bridge once carried both vehicular traffic and trains. The truss section is a lower deck for the railroad crossing. The main deck is supported on the tops of the arches. One of the arches was destroyed by the German army in 1944, but it was later rebuilt and widened.

Vital statistics

Each arch has a span of 617 feet
and rises over 89 feet, giving them
a span-to-rise ratio of 7:1. The total
length of the bridge is 2,913 feet.

Main deck (right)

Five parallel reinforced-concrete beams
support the bridge's upper deck along
the approach section. Note the flange
(ridge) between the outside of the external
beam and the deck, which transfers the
loads from the overhanging 30-foot-wide
deck into the beam.

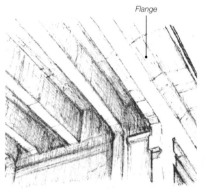

Flange

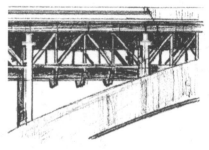

Hollow box

The route of the lower deck appears to
be blocked by the crown of the arch, but
it does, in fact, pass through the structure
because it is hollow. The hollow box arch
was constructed using a floating section of
falsework. Once the concrete has acquired
an appropriate strength as it cures, the
falsework can be removed.

Double deck

The series of trusses beneath the main deck
create a secondary deck beneath. Both the
main-deck concrete structure and the steel
trusses beneath span between the slender,
vertical concrete piers in the open spandrel,
which transfers the load into the arch. Note
the transverse bracing beneath the railroad
deck, which stiffens it between piers.

Salginatobel Bridge

Economical design
Robert Maillart's competition-winning design for the Salginatobel Bridge (1930) was successful not for its sleek form but for its economy. It was the cheapest design among all the entries.

Robert Maillart's 436-foot-long bridge (1930) across the Salgina Valley in Switzerland heralded a revolution in the use of reinforced concrete in bridge construction. The bridge is renowned for the economical way its reinforced-concrete deck bears heavy loads over its 295-foot span. Concrete hinges were built into the structure at the foundations and at the crown of the arch, creating a three-pinned structure. The bridge has a slight gradient, with a height difference between both ends of around 13 feet.

Three-pinned arch

Three-pinned arches have several
advantages over two-pinned or fixed
arches, in which horizontal thrusts and
bending moments at the center are difficult
to determine. A three-pinned structure
can be calculated accurately. It is also
better at accommodating movement,
caused either during the construction
process or by expansion, contraction,
or creep during its working life.

Structural efficiency (below)

The individual elements of the arch reveal
the structure's efficiency. The arch increases
in width at the abutment to dissipate the
loads into the foundations. The walls are
inset from the arch and deck and increase
in height to the quarter span, reflecting the
bending-moment distribution. The evenly
spaced columns supporting the deck are
linked transversely by slender slabs that
stiffen the structure.

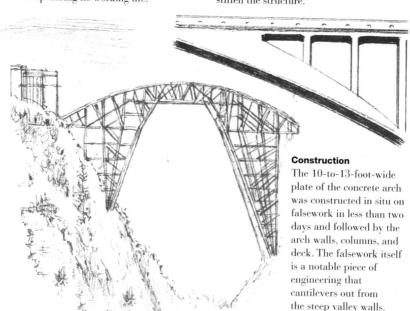

Construction

The 10-to-13-foot-wide
plate of the concrete arch
was constructed in situ on
falsework in less than two
days and followed by the
arch walls, columns, and
deck. The falsework itself
is a notable piece of
engineering that
cantilevers out from
the steep valley walls.

Sydney Harbour Bridge

Arch thickness

The thickness of the Sydney Harbour Bridge arches varies from 59 feet at the crown, where the loads are least, to 187 feet at the approach-road end, where the loads are greatest.

Standing 440 feet above sea level and with a span of 1,650 feet, Sydney Harbour Bridge (1932) is still among the world's tallest and widest steel arch bridges. Fabricated from 58,000 tons of steel and more than six million rivets, the two-pinned-arch structure is formed of two truss arches joined by crossed bracing. The deck is suspended from the arch on hangers that descend from beneath the vertical elements of each truss section. The width of the deck is substantially wider than the arch structure and carries eight lanes of traffic, two railroad tracks, a cycle lane, and a footpath.

Through arch

The deck of the bridge intersects the arch above the abutments, making it a through-arch bridge. Note how the granite-clad concrete pylons are detached from the arch structure except at their base and, therefore, serve no specific structural function.

Half-arch construction

Each side of the arch was built up from the abutments using electrically powered cranes that crawled up the arch as it grew. To prevent the half arches from toppling during construction, they were held back by 128 steel cables anchored into the ground. The arch is an example of the Pratt-truss arrangement, where the diagonal members in each of the twenty-eight truss sections slope down toward the center of the arch.

Abutments

The foundations were dug to a depth of 40 feet and excavated 4.31 million cubic feet of rock. They were filled with high-grade concrete to support the four massive pins on which the entire structure sits. Each pin is almost 14 feet wide and 14 inches thick, resisting a thrust of approximately 22,000 tons. The pins are also hinges that allow for the bridge to expand, contract, and rotate, which can alter the height of the bridge by up to 7 inches.

New River Gorge Bridge

Weathering

The New River Gorge Bridge was built using COR-TEN steel, a type of weathering steel that does not require painting. The result is a natural rusty-colored finish, in keeping with the natural environment in which the bridge is located.

When completed in 1977, the New River Gorge Bridge in West Virginia was the longest single-span arch bridge in the world and the highest arch over water. The two-pinned, steel-truss segmental arch has a span of 1,696 feet and rises to 876 feet. The arch supports a continuous truss deck with two expansion joints over the piers rising from the abutments. The steel piers supporting the deck are evenly spaced and rise from the valley sides and across the arch. The arch was constructed simultaneously from both sides, each side being anchored by steel wires until the arch met in the middle.

Arch section lengths (above)

The profile of the main arch is slightly tapered, being thicker at the base, where the loads are greatest, and thinner at the crown, where the loads are least. The piers are similarly tapered from the base to the top. The cross-braced supporting columns are connected to the arch at every third truss section. Note how these arch sections vary in length to ensure a uniform distance between the piers is maintained along the length of the bridge.

Foundations (above)

The foot of each arch comprises a steel-pinned connection to allow for rotation generated by loading and expansion and contraction of the steel. Note how the structure has been strengthened where it meets the pin joint by building up a series of riveted steel plates.

Deck truss

The road deck is constructed on a continuous through-type subdivided Warren-truss beam supported on the tapered columns at every third truss section. The deck is stiffened by lateral and longitudinal cross bracing underneath.

Bloukrans Bridge

Environmental impact

An arch bridge was selected to cross the Bloukrans River to minimize the impact on the surrounding natural environment. The rocky valley sides were ideal for supporting the thrust forces in the arch.

Towering 709 feet over the Bloukrans River in South Africa, the Bloukrans Bridge (1984) is the highest single-span arch bridge in Africa. The 892-foot-span, reinforced-concrete, deck-arch structure stands on foundations built into the valley walls at a height of 490 feet above the valley floor. Pairs of slender columns support the 1,480-foot-long and 52-foot-wide concrete deck. The arch has no hinges at its feet or crown, making it a fixed arch. The space between the crown of the arch and the deck accommodates the world's highest bungee jump (709 feet).

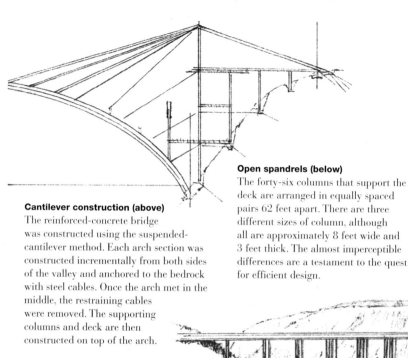

Cantilever construction (above)

The reinforced-concrete bridge was constructed using the suspended-cantilever method. Each arch section was constructed incrementally from both sides of the valley and anchored to the bedrock with steel cables. Once the arch met in the middle, the restraining cables were removed. The supporting columns and deck are then constructed on top of the arch.

Open spandrels (below)

The forty-six columns that support the deck are arranged in equally spaced pairs 62 feet apart. There are three different sizes of column, although all are approximately 8 feet wide and 3 feet thick. The almost imperceptible differences are a testament to the quest for efficient design.

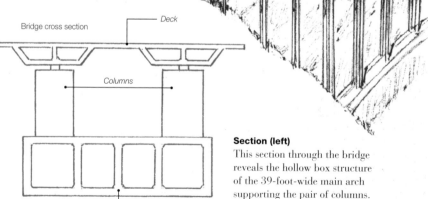

Bridge cross section

Deck

Columns

Main arch
cross section

Section (left)

This section through the bridge reveals the hollow box structure of the 39-foot-wide main arch supporting the pair of columns. Each 8-foot-wide column supports a 26-foot-wide hollow deck section.

Chaotianmen Bridge

Chords

The principal members of a truss are the chords, between which the vertical and diagonal members are arranged. Here, the upper and lower chords are red.

In 2009, the Chaotianmen Bridge across the Yangtze River in the city of Chongqing (the world's most populous city) in southwest China became the world's longest arch bridge. The main span of the half-through arch is 1,811 feet long and 466 feet high and carries a double deck. The bridge's main span is a two-pinned, steel-trussed arch supporting a pair of steel decks that are stiffened with lateral-transverse bracing.

朝天门长江大桥

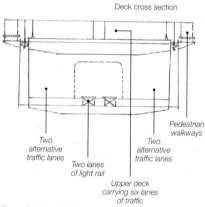

Truss arch

The bridge's wider upper deck extends beyond the sides of the arch. The continuous steel-truss structure can be seen passing through the deck. Note how the truss's diagonal members change direction where the arch meets the ground. This change of direction maintains tension in the diagonal members that carry the loads back to the support. The top and bottom members of the truss are fabricated from welded box-steel sections.

Double deck (below)

The 120-foot-wide top deck carries six lanes of road traffic and two pedestrian paths on each side. The 95-foot-wide lower deck carries two railroad tracks and four lanes of traffic. The upper deck is stiffened with longitudinal U-shape ribs and the lower deck, which is suspended from the upper deck by tapered steel columns, is braced on its underside with transverse and diagonal steel members.

Deck cross section

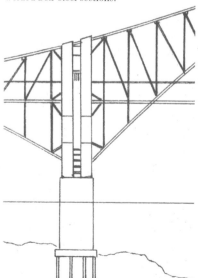

Pedestrian
walkways

Two
alternative
traffic lanes

Two lanes
of light rail

Two
alternative
traffic lanes

Upper deck
carrying six lanes
of traffic

Foundations (left)

Note how the foundations are required to take only a vertical load because the truss is tied. This is in contrast to abutment-type foundations for arch bridges, which need to resist horizontal thrust.

Introduction

Truss bridges rely on the inherent strength of the triangle, whose individual members are subject to compression, tension, or both (though not at the same time). The two principal elements of a truss are the two chords (upper and lower), the outside members of a truss that extend from one end of the truss to the other and are connected by a series of smaller vertical or diagonal members in compression or tension. There are many configurations of trusses, which create a wide range of different patterns

and appearances on bridges, from simple triangular types to very complex latticework. The amalgamation of truss elements into an entire structural framework performs the function of a beam, arch, or cantilever, and provides support for the bridge.

Truss bridges can, therefore, be as simple as a triangle carrying a footbridge over a stream or as complex as a series of cantilevered sections carrying a multideck road-and-rail bridge across the sea.

Kingston–Rhinecliff
The Kingston–Rhinecliff Bridge (1957) in New York comprises a series of different truss types that create ten spans and carry two lanes of traffic across the Hudson River.

Pont-y-Cafnau

Bridge of troughs
A column rose from the structurally rigid point of the Pont-y-Cafnau, where the apex of the truss met the parapet to support an overhead trough that once carried water and limestone to the ironworks.

Constructed in 1793, the Pont-y-Cafnau over the Taff River in Wales is an example of an early type of truss bridge and is the world's oldest surviving iron railroad bridge. Built to serve a local ironworks, the bridge once comprised two independent decks: a railroad at ground level and an aqueduct above. The 46-foot-wide deck is a cast-iron hollow box, 6½ feet wide and 2 feet deep, supported by three lateral beams connected to two large A-frame trusses on each side. The apex of each truss meets the top of the parapet in the middle of the span.

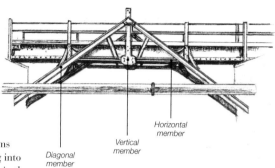

A-frame truss (right)

The horizontal tension member of the A-frame ties the two diagonal sections together at the same height as the bottom of the deck, with the lateral beams supporting the deck socketing into its sides. The vertical member in the center of the A-frame extends above the parapet to provide a bracket in which the wooden columns supporting the aqueduct were secured.

Diagonal member

Vertical member

Horizontal member

Carpentry in iron (below)

The bridge's designer, Watkin George, was a trained carpenter and planned the bridge as if it were to be fabricated in wood. The mortise-and-tenon and dovetail joints in the wrought-iron members reflect his former vocation.

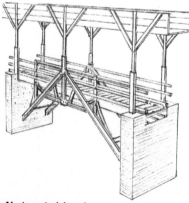

Abutments (above)

The valley walls were reinforced with vertical stone abutments. These support the vertical forces from the deck, which sits on top of the abutments, as well as lateral forces in the diagonal members of the A-frame lower down the abutments.

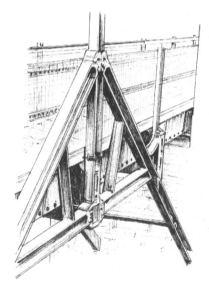

Busseau sur Creuse

Continuous beam
The railroad deck of the Busseau sur Creuse Viaduct is a continuous-lattice-truss beam. This is revealed by the absence of joints along the truss over the piers. A continuous truss minimizes the beam's depth and is more efficient than simply supported spans. The beam is supported by five piers and stone abutments at each end, one of which contains three round arches. Four of the six spans are 165 feet long, one is 147 feet, and another is 134 feet.

Lattice trusses were first patented in the early 19th century. The deck of the railroad viaduct in Busseau sur Creuse (1863), France, is laid on top of the beam, whose sides are constructed of closely spaced parallel rows of flat wrought-iron members riveted together at 45-degree angles. The piers stand on masonry foundations on the valley floor. The entire bridge is 1,109 feet long and weighs over 2,200 tons, including 16½ tons of paint.

Deck, pier, and railing

The 6-foot-deep lattice trusses not only form the walls of the deck, but are also replicated across the 26-foot width of the deck. Two vertical members strengthen the truss in the areas of high shear force, where it is supported on the piers.

Tapering piers

The tapered wrought-iron piers decrease in cross section as they rise from the foundations to the deck. The taper is more extreme in the direction perpendicular to the span. The piers are constructed in sections, each of which is defined by horizontal bands strengthened by diagonal bracing. Note the bolted flange details between sections. The masonry foundation supporting the tallest pier is 59 feet high and the height of the bridge at that point is 184 feet. The wrought-iron piers are flanged at their base to provide bolt housings, where they are connected to the masonry piers.

Howrah Bridge

Clearance

The Howrah Bridge's 100-foot-wide deck (with a 72-foot central section for traffic between two 14½-foot pedestrian lanes) has a clearance above water of 30 feet.

Truss bridges commonly support the deck on, within, or at the base of the truss, but the deck of the Howrah Bridge (1943) over the Hooghly River in Kolkata (Calcutta), India, is suspended beneath the large cantilever truss. Made from riveted steel members, the 22,000-ton structure comprises three main sections: two cantilevers and a suspended central section. Joints are located between these three sections to allow for thermal expansion and contraction. The main span is 1,500 feet and the towers of each cantilever section are 272 feet tall.

Deck (below)

The deck is suspended from the lower chord of the main structure by thirty-nine pairs of vertical steel-truss members. Compared with the main cantilever members, these "hangers" can be relatively slender because they are working in tension and not compression, so they are not prone to buckling.

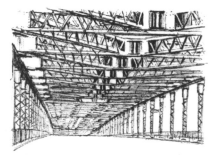

Anchor arm (below)

Because the deck is suspended beneath the truss, it reaches land at the tower, leaving the portion of the cantilever on the land side of the tower free from carrying deck loads. Note the K-truss diagonals are oriented in the opposite direction to the Minato Bridge (see page 151), therefore, the bottom members are acting in compression and the top ones in tension. The bottom diagonals are further braced to withstand buckling forces.

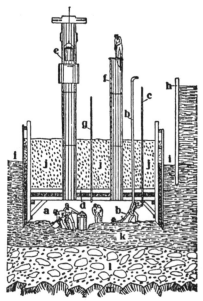

Caisson (above)

The Howrah Bridge has the largest land-based caisson in the world. Caissons are chambers lowered into the river and filled with air to create a space in which workers dig out the soil. The caisson sinks as the soil is removed until they reach a suitable substrate. The bridge's massive caisson was dug to a depth of over 95 feet. Five hundred people were employed to maintain the pressure inside the caisson to prevent it from filling with water.

TRUSS BRIDGES

Kingston–Rhinecliff Bridge

Cofferdams

Construction on the piers began in 1954. The foundations were constructed using cofferdams, in which large sheets of metal are placed into the riverbed to form an enclosed space. The water is pumped out and the concrete poured in to form the base of the piers.

Originally intended to be a suspension bridge, the bedrock on the banks of the Hudson River was found to be inadequate for anchoring the cables, so a bridge comprising a series of deck trusses was selected instead. Opened in 1957, the 33-foot-wide, two-lane steel-truss bridge has ten spans: four central spans of approximately 490 feet; two main spans of 800 feet; and four smaller spans partially on land. The bridge's entire length is just over 1⅓ miles.

Bearing joints (right)

Each of the nine piers between the main abutments is constructed of pairs of cross-braced, reinforced, tapered concrete columns. Two bearing joints on top of each pier support the trusses at its outer wall. Eight of the piers stand on deep foundations in the river and one is on land.

Navigation channel (above)

The arched profile of the main navigation channels reveals their longer spans. The truss is deepest at its ends, where the shear forces are greatest, and the profile causes the truss to act as a cantilever over these supports. The deck is supported on top of the truss along the entire length of the bridge.

Subdivided Warren truss (below)

Each section of the bridge is constructed using a subdivided Warren truss. Some sections are simple truss beams and others are cantilever trusses. The joint between truss sections above some of the piers reveals that the bridge is not a continuous truss over its full length and allows for movement between the bridge sections.

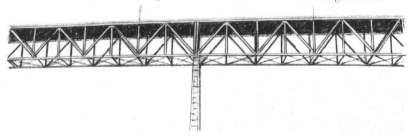

Astoria–Megler Bridge

Engineered safety
The Astoria–Megler Bridge is exposed to harsh natural conditions. This truss bridge is designed to withstand violent ocean storms, and its steel piers stand on elongated concrete footings to protect them from debris carried downstream by floodwaters.

The main section of the 4-mile-long Astoria–Megler Bridge (1966) in Washington is the longest continuous-truss bridge in North America. The main truss, which has a different height at each end, creates a navigation channel in the Columbia River with a maximum clearance of 197 feet and a span of 1,234 feet. The approach spans are beam trusses of a subdivided Warren type. Most of the bridge is constructed using low-lying, simply supported prestressed-concrete beams.

Camelback truss (above)

The entire bridge system comprises a series of different truss structures. On the Washington shore side, the bridge is made up of seven simply supported subdivided Warren-type, camelback through trusses. The name camelback comes from the distinctive humpbacked profile of the polygonal upper chord.

Ramps (below)

One approach to the continuous-truss section is a continuous-beam truss supported on five concrete piers. The steep gradient of the approach ramp from the high continuous truss down to the low section fabricated from prestressed-concrete beams requires some horizontal restraint to secure the deck on support piers.

Tapered piers (right)

K-braced tapered piers provide both lateral and longitudinal support to the continuous-truss section. The sharp angle of the taper provides a wide supporting base on a concrete foundation that rises to a knife-edge apex supporting the truss's load and stabilizing it against any lateral forces, such as wind loads. The knife-edge support enables the truss to rotate about this point in response to loads and expansion/contraction movements.

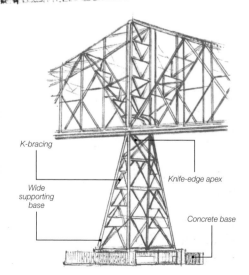

K-bracing

Knife-edge apex

Wide supporting base

Concrete base

Minato Bridge

Through truss

The 72-foot-wide double deck is housed inside the truss beneath and parallel to the upper chord, making the Minato Bridge a through-truss bridge. The 40,000-ton structure has the third longest span of any cantilever bridge in the world.

The Minato Bridge (1974) is a double-deck steel cantilever through-truss bridge in Osaka, Japan. The two large cantilever sections are anchored on piers on the shoreline and extend over the water to create a span 1,673 feet long and 167 feet high. The structure of the cantilever sections are K-trusses, evolving into a simple Pratt truss in the suspended-truss section in the center of the span. A cantilever truss was considered the most appropriate configuration for this particular bridge because the poor quality of the subsoil was unable to support the weight of an arch bridge.

K-truss

The K-truss sections are clearly visible
on each side of the central part of the
cantilever. Note how the K is oriented
in the direction of the cantilever. This
means that the lower leg of the K always
works in tension and the upper works in
compression. Note also how the central part
of the K remains at a consistent height
in line with the bridge's lower deck and,
therefore, decreases in height as
the truss section narrows,
turning into a Pratt
truss at the bridge's
narrowest central
portion. (See also the
Howrah Bridge,
pages 144–145.)

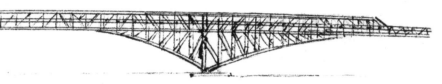

Not an arch (above)
Viewed in profile, the two
cantilevered sections of the
Minato Bridge reveal the length
of span that can be achieved
compared with an arch, which
would have required a much
larger structure. The entire
length of the bridge, including
truss-beam approaches,
is 3,225 feet.

Lateral ties (above)
Lateral Warren-type trusses tie both sides
of the bridge together, strengthening the
structure and preventing it from twisting.

Ikitsuki Bridge

Renewable energy
The velocity of the water current is greater around the Ikitsuki Bridge piers, making them ideal sites for generating renewable energy.

With a main span of 1,312 feet, the Ikitsuki Bridge (1991) is the longest continuous-truss bridge in the world. Truss bridges with longer spans than the Ikitsuki Bridge are not constructed from a continuous truss but have cantilevered sections supporting a suspended midsection. The truss sections are a combination of K-trusses and a Pratt truss in the midsection.

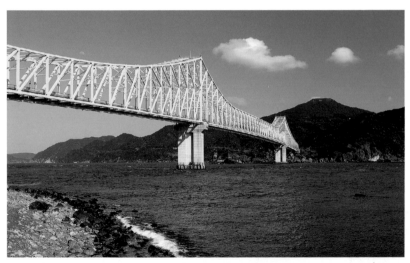

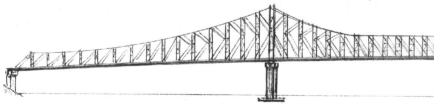

Through truss (right)

The road approaches the bridge at one end over a short section of concrete beams simply supported on concrete piers before passing through the truss structure, making it a through truss.

Piers (above)

Two concrete piers in the open water between Ikitsuki and Hirado Island in Japan support the bridge at the point where the truss profile rises. Each pier comprises two supporting legs joined by a thin wall of concrete and connected at their top by a lateral load-bearing beam. The corbeled profile of the beam reflects the way in which it cantilevers beyond the piers to restrain laterally the outer edges of the bridge deck.

Continuous truss (below)

A continuous truss is defined as a spanning-beam structure that develops hogging moments over the supports and sagging moments in the center span. Although it appears similar to a cantilever bridge and might have been constructed as two cantilevered sections temporarily, it does not behave like a cantilever. The principal difference is that a continuous-truss bridge relies on the structural performance of the central span as part of an integrated structural form.

Ponte della Costituzione

Designed by Santiago Calatrava and opened in 2008, the Ponte della Costituzione over Venice's Grand Canal is a steel-arch truss fabricated from five arch sections: a central arch, two side arches, and two lower arches. The five tubular-steel arches are connected to one another with perpendicular tubular and plate-steel members, creating a complex skeleton structure of ribs along five spines. The central and lower arches form a main spine, which is the principal supporting element, while the two pairs of outer arches provide additional support to the splayed deck.

Quiet revolution (left)

Before this bridge was built, none of Venice's bridges exceeded a span-to-rise ratio of 7:1. With a span of 66 feet and a rise of just 15½ feet creating a span-to-rise ratio of 16:1, the Ponte della Costituzione has instigated a quiet revolution in the fundamentals of bridge design in Venice.

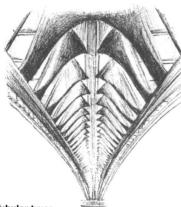

Tubular truss

The smaller transverse members, or ribs, are angled radially, pointing down away from the center of the bridge's span. The absence of any longitudinal triangulation in the truss (that is, a combination of verticals and diagonals) makes this a Vierendeel truss.

Clearance (below)

The single-span arch has a 590-foot radius, creating a shallow arc with a span of 262 feet and a 23-foot clearance. Making the arch stiff enough to work in tension, and not just compression, has resolved two competing needs: pedestrians passing over avoid a steep climb, and boats passing under are not obstructed.

Foundation

The abutments are constructed from reinforced concrete, which is clad in stone. The identical sculpted forms have been carefully designed to complement the gentle arc of the bridge as it touches down on either side of the canal.

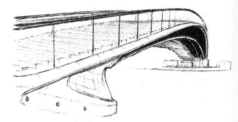

Glass deck

The gently rising deck, comprising glass steps with lights within each step, increases in width from the abutments to the midspan. A parapet of transparent plate glass topped with a bronze handrail maintains the bridge's lightweight and slender appearance.

OPENING & MOVING

Introduction

Moving bridges provide a means of crossing that can be temporarily removed. The most common function of a moving bridge is to allow for tall boats to pass, but rarer examples include defensive functions, such as drawbridges, transportive functions to move goods across a river, or the protective function of removing the structure from the path of a flooded river. Opening bridges are not confined to any particular structural type, such as arch or beam, and come in all shapes and sizes. More than any other

category of bridge, moving bridges present a unique opportunity for design innovation, making them among the most flamboyant bridge type. Moving bridges often also present complex engineering challenges, because their movable elements exert shifting and reversible forces on the structure. Among the many types of moving bridges around the world, there are those that ascend and those that descend, while others swivel, and there are even some that curl or tilt.

Gateshead Millennium Bridge
Completed in 2001, the multiaward-winning Gateshead Millennium Bridge across the Tyne River in northeast England is the world's first and only tilting bridge.

Barton Swing Aqueduct

Cross bracing
The cross bracing in the central section of the truss over the pivot of the Barton Swing Aqueduct provides additional strength over the support. The aqueduct swings at its midsection on a specially constructed island in the middle of the canal.

The Bridgewater Canal in northwest England was constructed in the mid-17th century to transport coal to Manchester. As one of the earliest transport arteries built in response to Britain's rapid industrialization, it had to successfully traverse existing natural waterways and established transport routes. A stone aqueduct was built for the purpose of crossing the Irwell River, which was replaced in the late 19th century by a swing aqueduct—the first and only one of its kind in the world when it opened in 1894. The replacement was necessitated by the construction of the Manchester Ship Canal in the 1890s, along which much larger ships traveled up the waterway beneath the original aqueduct.

Swing pivot (above)

The 1,600-ton movable section of the aqueduct is a 233-foot-long and 23-foot-wide steel through truss. Note that the truss arrangement looks like a typical simply supported Howe truss. However, the diagonals are in tension, not compression, because of the cantilever form.

Twin pivot

The aqueduct is not the only swing bridge at this point of the Manchester Ship Canal. An adjacent swing bridge carrying a road over the canal shares the artificial island. The bridges are designed to work concurrently so that in their open positions they are aligned on the island, separated only by the brick control tower.

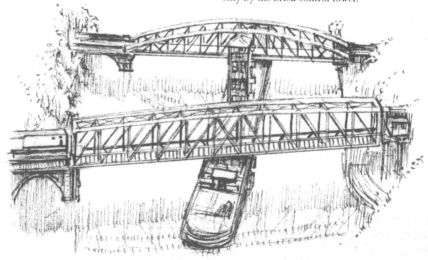

Tower Bridge

Why the need?

Tower Bridge was designed to let tall ships reach the Pool of London between Tower Bridge and London Bridge. Clearance when the bascules are closed is 26 feet and 138 feet when open.

One of the most famous moving bridges in the world crosses the Thames just outside the eastern boundary of the old City in London. Tower Bridge, so named for its proximity to the Tower of London on the northern bank, is a combination of a bascule and a suspension bridge and was opened in 1894. The mechanisms for the bascules are housed at the base of the towers, which stand on piers made with 77,000 tons of concrete. Following its completion, it was the biggest and most sophisticated bascule bridge in the world.

Bascule and suspension

The 800-foot-long bridge can be divided into two structural forms: the 200-foot central span, with its pair of bascules, and two suspended outer sections. The outer sections comprise a deck that is supported by rods hung from suspension trusses. The horizontal tension forces from the suspended section are carried through a pair of lattice-truss box-girder beams between the main towers. These beams also house walkways.

Steel truss (left)

Although Tower Bridge has been covered in Portland stone and granite, most of the structural elements are steel and concealed inside. Over 12,000 tons of steel were used in the steel frame of the towers, the lattice trusses across the top, and the suspension rods.

Bascule (right)

Each bascule weighs more than 1,100 tons. To make the operation of raising them easier and to relieve the stress on the pivot, each bascule is counterbalanced. The counterbalance and the modern hydraulic mechanism used to open the bridge are housed within the piers.

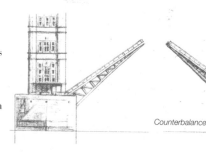

Counterbalance —

Middlesbrough Transporter Bridge

Transporter bridges allow for low-level navigation across a river while maintaining high-level clearance. This dual function is achieved by having a movable car, or "gondola," suspended on steel cables from a raised beam. The car travels back and forth across the river using cables and pulleys. The transience and portability of the bridge deck allows for the easy and regular passage of large boats under the framing structure while maintaining river crossing. The bridge is suited to sections of river where priority is given to waterborne traffic.

Middlesbrough Transporter

The Middlesbrough Transporter Bridge in northeast England opened in 1911 and is the last bridge across the Tees River before it reaches the sea. A transporter bridge was chosen because it did not restrict river navigation. Only twenty transporter bridges have ever been built, and only eleven of these survive.

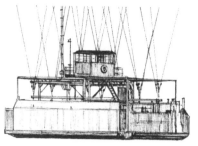

Transporter car
The Middlesbrough Transporter can carry 200 people or nine cars across the river in ninety seconds.

Cantilever (below)
Transporter bridges rely on a high-level structure supporting the cable mechanism from which the car is suspended. Many different methods can be used to achieve this, but the Middlesbrough Transporter uses two 161-foot-tall, steel cantilevered sections supported on tapered pylons that provide a central span of 590 feet. The cantilever arms are longest across the river and the shorter anchor arms over land are tied to the ground with cables to resist the uplift forces.

Truss piers (above)
The steel-truss piers supporting the high-level cantilevered sections are tapered asymmetrically so that the inner walls are vertical to allow for unrestricted passage of the cables and car, which is an important design consideration.

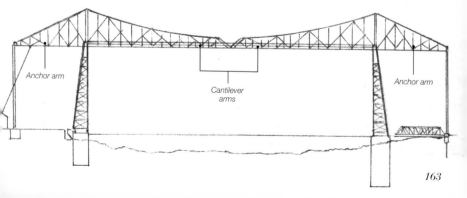

Anchor arm

Cantilever arms

Anchor arm

Michigan Avenue Bridge

Renamed

The Michigan Avenue Bridge was renamed the DuSable Bridge in 2010, after Jean Baptiste Pointe DuSable, the first nonnative settler in Chicago.

Designed to link the northern and southern parts of Chicago across the Chicago River as part of an ambitious early 20th-century urban plan for the city, the Michigan Avenue Bridge (1920) was the first double-deck bascule bridge ever built. Faster noncommercial traffic used the top deck, and slower commercial traffic used the bottom. Pedestrian walkways flank both decks. The bridge comprises two equal bascules, or leaves. Four tender houses stand on each corner of the stone abutments.

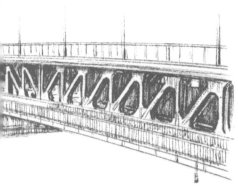

Deck truss

The structural form of the double deck in its lowered position resembles a Pratt truss. It functions as both a through truss (with the bottom deck) and a deck truss (with the top deck). Note that the trusses still function in a similar way to a simply supported Pratt truss, with the diagonals in tension, because the cantilever supports are at the ends and not the middle, like the Barton Swing Aqueduct (see pages 158–159).

Twin bridges (right)

Each leaf is divided longitudinally so that each side of the bridge can operate independently of the other in the event of one side being struck and damaged by a boat. Each leaf comprises twelve longitudinal steel beams strengthened by diagonal and horizontal cross bracing.

Trunnion bascule

The pivot of each 3,700-ton leaf is beneath the lower deck and counterweighted by four 1,600-ton weights that sink into 39-foot-deep pits as the bridge opens. The bridge pivots on cylindrical pin joints, or trunnions, allowing for it to move up and down.

Corinth Canal Bridge

Fishing bridge
As the bridge is raised, fish get caught on the deck in the space between the balustrades. The trapped fish are often collected by local children.

The submersible bridge is a comparatively rare bridge type. The method is similar to that of a vertical-lift bridge, but instead of the deck being lifted vertically, the deck of a submersible bridge descends to give enough draft for vessels. Both ends of the 3¾-mile waterway that forms the Corinth Canal between the Corinth and Saronic Gulfs in Greece have submersible bridges (1988) that descend to the bottom of the canal to let boats pass.

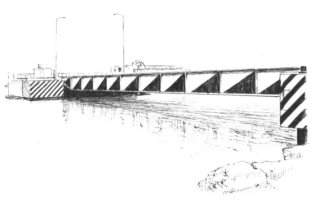

Bridge up

When completely raised, the bridge stands nearly 6½ feet above sea level with only a narrow clearance thereby forming an effective barrier to the canal and preventing any boats from entering.

Bridge submersing

The main deck comprises four 3-foot-deep steel girders supporting two lanes of traffic. Treatment of the steel is a critical design consideration to protect it from the aggressive marine environment. Both ends of the bridges are secured into reinforced concrete piers that house the rising mechanism.

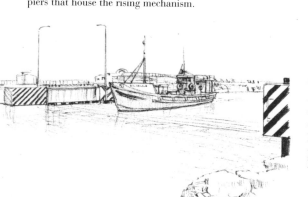

Completely submersed

The bridge maintains a horizontal position as it descends to the bottom of the canal, offering boats the maximum possible clearance beneath their keel.

Erasmus Bridge

Multiple piers
A number of different concrete piers support the 2,651-foot-long steel deck of the Erasmus Bridge, which carries vehicular traffic and urban trams, cyclists, and pedestrians.

A moving section forms a small portion of the prominent Erasmus Bridge over New Meuse River in Rotterdam, The Netherlands. The 292-foot bascule is dwarfed by the 920-foot main span suspended by cables from a 456-foot-high pylon, but it is a vital part of the bridge because it is the only section through which large boats can pass. Opened in 1996, the Erasmus Bridge is the tallest bridge in The Netherlands, and its bascule section is the largest and heaviest bascule bridge in Western Europe.

Bascule (below)

The structure of the bridge's large bascule is tapered so that it is thicker near the pivot and thinner toward the outer edge. The stronger cross section close to the support, or pivot point, reflects the requirements of the cantilever from. The bascule comprises two principal cantilever arms connected by transverse beams over which the deck is laid.

Tower cable (above)

Nearly 7,700 tons of steel were used in the construction of the Erasmus Bridge. Much of this was used in the angled pointed pylon that supports the thirty-two cable stays, the longest of which are nearly 980 feet.

Pylon (below)

The angle of the pylon connected to the horizontal base section of the bridge resists the loads applied by the deck and is tied to the rear of the bridge by eight anchoring stays. Note how it leans against the tensile deck forces.

Gateshead Millennium Bridge

Award winner
As the world's first and only tilting bridge, the Gateshead Millennium Bridge has won many international awards for architecture.

Crossing the Tyne River and opened to the public in 2001, the Gateshead Millennium Bridge was the world's first tilting bridge. Designed for pedestrians and cyclists crossing between Newcastle and Gateshead in northeast England, the bridge comprises two counterbalanced steel arches with a span of 345 feet. One of the arches forms the deck and is connected by steel cables to a supporting arch that acts as a counterweight to assist the tilting action.

Pivots (below)

Eight electrical motors drive the hydraulic mechanism that tilts the bridge over 40 degrees in less than five minutes. The bridge is also designed to be self-cleaning. Any litter left on the bridge when it is tilting rolls into traps at the end of the arch.

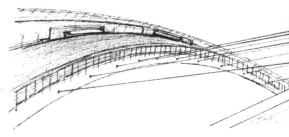

Deck cables (above)

The deck is divided into two sections. The cycle lane is on the outer curve and is about 1 foot lower than the pedestrian deck on the inner curve. The steel cables connect to the inner edge of the deck arch to avoid obstructing head height clearance for pedestrians and cyclists. This means that the deck is cantilevering out from these lines of support.

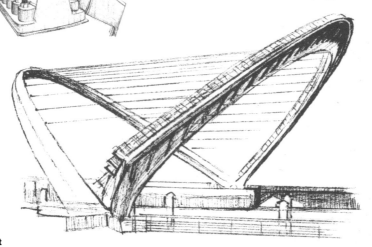

Tilt

To allow for boats to pass, the 880-ton steel structure tilts from pivot points on abutments constructed from 21,000 tons of concrete that project from each bank. As the supporting arch lowers, the deck arch rises until both are equal in height.

The Rolling Bridge

The 39-foot-long pedestrian Rolling Bridge (2004) over an inlet of the Grand Union Canal in London comprises eight hinged deck sections with polygonal walls that curl up to form a perfect octagonal structure on one bank. In its unfolded form the bridge is a truss—the handrails forming the upper chord and the deck the lower chord. The slight downward angle of the handrails transfers the loads in compression into the vertical cylindrical posts. Folding takes three minutes and can be stopped at any point during the process.

Operation

At the touch of a button, the bridge begins to curl as it rises to a height of almost 33 feet before coiling into an octagonal form on one side of the inlet in under three minutes.

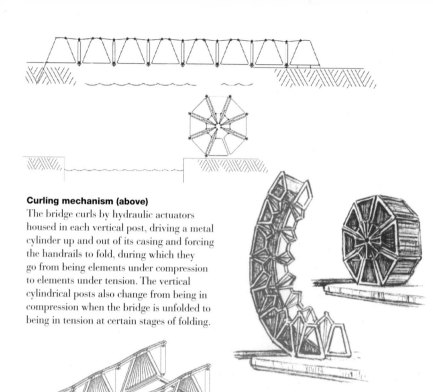

Curling mechanism (above)

The bridge curls by hydraulic actuators housed in each vertical post, driving a metal cylinder up and out of its casing and forcing the handrails to fold, during which they go from being elements under compression to elements under tension. The vertical cylindrical posts also change from being in compression when the bridge is unfolded to being in tension at certain stages of folding.

Folding sections

The principal elements of the bridge are the polygonal balustrade panels, the vertical cylindrical posts, the hinged U-section handrails (the sides of the U providing lateral support to the vertical posts), and the deck sections. To keep the weight of the bridge to a minimum, the voids in the balustrade panels are filled with wires.

Curling sequence

As the bridge starts to curl, it transforms from a simply supported truss to a cantilevered truss. In the initial phase of curling, as the handrails pass through horizontal, the structure arches. A hinge in the fixed deck section allows for this movement. The bridge rises to its full height then coils into an octagonal structure.

Gustave-Flaubert Bridge

A vertical-lift bridge comprises two pylons that support both ends of the deck and raise it in a vertical motion while keeping the deck horizontal. One of the largest vertical-lift bridges in Europe is the Gustave-Flaubert Bridge (2008), the sixth bridge to be built across the Seine River in Rouen, France. The total length of the bridge, including approach roads, is 2,200 feet.

Rising system
The 381-foot central span of the Gustave-Flaubert Bridge carries twin road decks fabricated from steel girders. These are raised and lowered on a series of steel cables strung over 500-ton pulley mechanisms, which are housed at the top of each pylon.

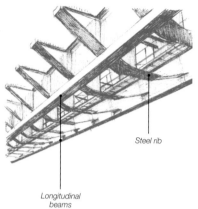

Steel rib

Longitudinal beams

Road deck

Each 381-foot-long steel road deck weighs 1,300 tons. The main structure comprises two longitudinal beams that are connected and strengthened by a series of steel ribs, which cantilever from the sides to produce a suitable deck width.

Pylons

The pair of 282-foot-high reinforced-concrete pylons each comprises two independent masts joined at the top by the pulley mechanism. This produces structural depth across the pylons by ensuring that they work together. The road decks on each side of the pylons are suspended at both ends by two pairs of cables—one pair held to the outside of the deck and the other attached to the inside of the deck.

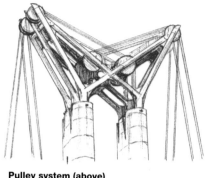

Pulley system (above)

The butterfly pulley mechanism at the crown of each pylon houses eight winches. Note how the structure of the mechanism is strengthened by being cross braced. More than 3¾ miles of steel cable is wound around each mechanism to raise the road deck 157 feet in twelve minutes.

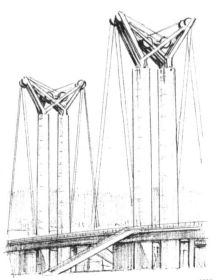

Cathedral Bridge

Historic reference

The piercing form of the Cathedral Bridge's 66-foot pylon evokes a scissor or needle, recalling the world of textile manufacturing that made Derby famous. The swinging action of the 197-foot-long bridge also suggests the cutting action of tailor's shears.

Situated amid a UNESCO World Heritage Site, the new pedestrian bridge over the Derwent River in Derby, England, required a sensitive and unobtrusive design. The pedestrian and cycle bridge links the eastern bank of the river with the Cathedral Green, the historic town center, and the famous Silk Mill Museum, where the modern factory was born. The new Cathedral Bridge (2009) is a hollow box-steel cable-stayed pedestrian swing bridge that is designed to rotate swiftly in response to the river's fast-flowing and rapidly rising floodwaters. The unobtrusive and slender deck is supported by three steel cables connected to a distinctive pointed pylon.

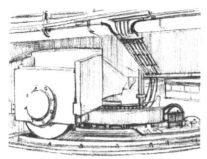

Pivot

The bridge's swinging mechanism is housed beneath the bridge under the mast at the point where the deck kinks at an angle of 38 degrees. The huge cast-steel pivot bearing moves the 99-ton structure in four minutes.

Swing motion

The counterbalanced bridge is designed to swing very efficiently about a vertical axis. The efficient design allows for the bridge to be opened using an electrical motor or operated by hand.

Environmental design

The kinked deck is perfectly balanced to rotate, and its thin bladelike profile helps to increase maximum clearance for floodwaters. The lean structure also reduces materials, all of which were sourced and fabricated within 15 miles of the site.

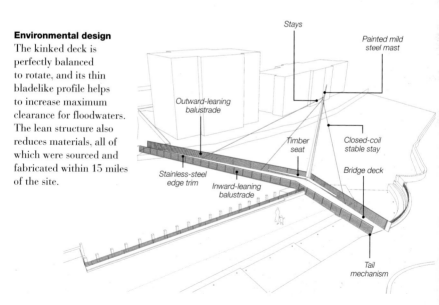

Stays

Painted mild steel mast

Outward-leaning balustrade

Timber seat

Closed-coil stable stay

Bridge deck

Stainless-steel edge trim

Inward-leaning balustrade

Tail mechanism

Introduction

A cantilever is a structure projecting from a support. In bridge design, the cantilever is one of the four basic structural types. Cantilever bridges are usually balanced on both sides of a support with an anchor (or back) arm forming the back span and opposing the cantilever arm, which forms part of the main span. Compressive forces on the underside of each arm counter the tensile forces in the upper part and are transferred into the support. Most cantilever bridges comprise at least a pair of cantilevers that meet to form the main span.

Cantilever bridges can be constructed from all types of materials, using different structures for the cantilever section, such as a beam or a truss. Balancing over a supporting pier, the cantilever arms are either secured to abutments or to adjacent cantilever sections. A suspended section is sometimes inserted between the cantilever arms to extend the main span. To counter the additional vertical loads to the main span, the balancing arms are anchored either to adjacent cantilevers or to solid abutments in the ground at each end of the bridge.

Quebec Bridge
The cantilever bridge with the longest span in the world is the Quebec Bridge (1919), in Canada. The 1,800-foot span is due to the extended suspended-truss section between two balanced cantilevers.

Forth Rail Bridge

Chords

The upper and lower chords of each cantilever of the Forth Rail Bridge are fabricated from tubular steel and transfer the loads back to the piers. The anchor arms are weighted with 1,100-ton weights to counterbalance the suspended section and live loads.

Among the most famous cantilever bridges in the world is the railroad bridge built over the Firth of Forth in Scotland (1890). The Forth Rail Bridge was the largest execution of the cantilever principle ever attempted in bridge construction. The bridge comprises three enormous, balanced double-cantilever sections joined together by small, simply supported sections that create two central 1,709-foot spans. The two back spans at each end are 659 feet long and connect to masonry pylons supporting the approach sections, which are fabricated from steel-truss beams.

Human cantilever

A human demonstration was carried out to illustrate the structural theory of the cantilever. The loads are represented by the figure sitting in the middle span on the suspended section. The arms of those seated on each side represent the tension in the ties. The wooden struts represent the compression in the lower members. The bricks represent the anchor points at the pylons.

Riveting

This was the first major bridge built entirely of steel. The bridge construction employed more than 4,000 workers a day, 6.5 million steel rivets, and 72,000 tons of steel. The tubular lower members shown here are made from riveted sections of sheet steel to form a hollow structural tube resisting compressive forces from above and transferring them to the piers.

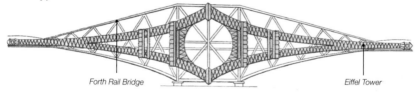

Forth Rail Bridge   *Eiffel Tower*

Grand designs (above)

It would take more than five Eiffel Towers laid end to end to span the same distance as the Forth Rail Bridge's three double-cantilevered sections.

Suspended section

A benefit of the balanced-cantilever method is the extra span gained by inserting a suspended section between each cantilever arm. The suspended sections on the Forth Rail Bridge are 348 feet long.

CANTILEVER BRIDGES

Quebec Bridge

Largest span

Although the cantilever arms of Quebec Bridge are shorter (581 feet) than the Forth Bridge, the central section is longer, at 640 feet compared with 348 feet, making the main span 1,800 feet.

The Quebec Bridge (1919) in Canada has the longest span of any cantilever bridge in the world. The first attempt to build the bridge ended in catastrophe because the structure collapsed in 1907, forcing a major redesign. The main element of the 3,238-foot-long riveted-steel structure is a through truss. Bearing a resemblance to the earlier Forth Rail Bridge (see pages 180–181), the Quebec Bridge has two balanced-cantilever sections linked by a suspended section forming the main span. The 95-foot-wide deck carries rail, vehicular, and pedestrian traffic.

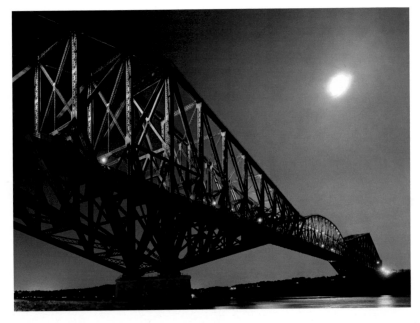

Center span

The 640-foot central section is a simply supported camelback truss. The entire central span was hoisted into place between the cantilever arms and connected to the bottom chord. The top chord infill element was installed to complete the structure. The outer arms of the cantilever are anchored to the abutments to counter the loads applied by the weight of the central section and the traffic passing over the bridge.

Cross bracing (left)

Both side faces of the bridge are tied together by a series of diagonal, vertical, and horizontal steel braces. Each of these bracing members is a truss element. All are elevated high above the deck, which also ties both sides of the bridge together.

K-truss (below)

The three separate structural elements of the bridge (both balanced cantilevers and the central section) are all types of K-trusses. Note how the bottom diagonals pointing back toward the supports are in compression and have additional bracing to withstand buckling forces.

Montrose Bridge

Suspended section
The Montrose Bridge's central span was extended by a 21-foot-long suspended section, which prevents a catenary (natural) curve forming in the top chord that is characteristic of a suspension bridge.

The curved chords that descend from the towers of the Montrose Bridge (1930–2004) over the South Esk River in Scotland give the impression that this is a suspension bridge. However, on closer inspection the concrete structure is created by two double cantilevers supported on two pairs of concrete piers, forming a main central span with two shorter side spans. The bridge replaced a suspension bridge, using its predecessor's abutments and approaches. The bridge was demolished and replaced in 2004.

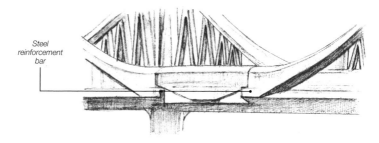

Steel
reinforcement
bar

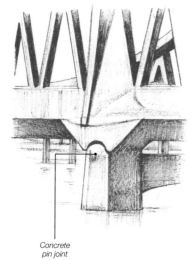

Concrete
pin joint

Chords (above)

The curved top chord contained seventy-six steel reinforcement bars, each approximately 1½ inches in diameter. The high ratio of steel to concrete in a member that is under tension fueled much debate about whether concrete was necessary at all, and, indeed, whether it was an appropriate choice of construction material for the structural form with regard to cost and weight.

Concrete sculpture (left)

An angular, sculptural concrete pin joint emphasized the meeting of the reinforced-concrete piers with the deck and superstructure.

Elevation

The two pairs of towers stood on reinforced-concrete piers with 59-foot-deep piles. The cantilevers, which were formed by concrete trusses with diagonal members between the curved top chord and the deck, created a main span of 217 feet. The shorter side spans anchored to the abutments were 151 feet long.

Story Bridge

Three piers

Uncharacteristically, Story Bridge stands on three piers. At the southern end, a main pier supports the cantilever and a secondary pier anchors the structure to prevent rotation. On the northern end, the cantilever stands on a main pier but is anchored in the bedrock.

Completed in 1940, the Story Bridge in Brisbane is the largest cantilever bridge in Australia. The steel cantilevers are constructed using a through truss, producing a 925-foot central span and a clearance of 98 feet over the Brisbane River. Construction of the steel structure began with the pylons standing on the concrete piers. The back-span arms were erected and anchored to masonry piers and approach roads before the structure cantilevered out into the central span. The inner arms, having reached their maximum length over the river, were joined with a simply supported central section completing the bridge's span.

Asymmetrical cantilevers

The outer cantilever arms anchored to masonry piers on land are shorter than the inner arms, which are suspended over the river. This creates an imbalance of load that must be restrained in tension at the anchorage points. The central section joining the two inner arms is a simply supported Pratt truss.

Cross bracing

The outer sides of the bridge are supported directly on the masonry piers and are cross braced with lateral and diagonal steel-trussed elements. The deck is laid above the cross bracing.

Through truss (above)

The bridge's K-braced steel structure forms a through truss with the road deck sitting on the base of the truss. The structure is supported on masonry piers, whose concrete foundations descend 130 feet.

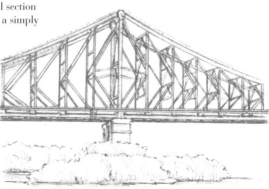

London Bridge

Long history
Over the last two millennia, many bridges have crossed the Thames River near the current London Bridge. The largest was a nineteen-arch, 12th-century bridge supporting tall buildings, a chapel, and a drawbridge.

Opened in 1973, the present 928-foot-long London Bridge stands near the site of London's earliest bridge, built by the Romans at a time when this part of the Thames River was the tidal limit. Because of the gentle curve of this cantilevered bridge, it is often mistaken for an arch bridge. However, a number of clues reveal the true nature of this cantilever structure—the two back spans are only half-arches and two movement joints are located in the central span. Four box beams fabricated from precast posttensioned-concrete segments form each cantilever section.

Movement joint (below)
Two movement joints are faintly visible in the central span. These indicate the suspended section between the two cantilever arms.

Posttensioned box beams (above)
Construction was complicated by the dismantling of the previous London Bridge (which was shipped to Arizona), which took place simultaneously while keeping the waterway open to river traffic. The four longitudinal box beams were prefabricated in segments downstream and shipped to the site, where they were connected to one another and posttensioned. Posttensioning involved inserting steel cables through each segment and tensioning the cables to assist and improve the structural performance of the concrete.

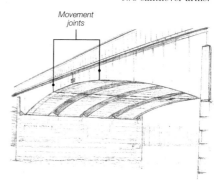

Movement joints

Half-arch (below)
The narrower side spans are formed by the outer arms of the cantilever sections, which spring from the slender granite-faced concrete piers. These cantilevers are anchored at the abutments.

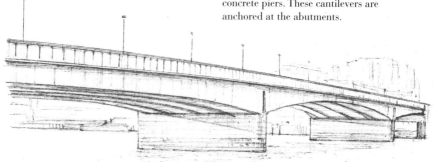

Commodore Barry Bridge

Old technology

Commodore Barry Bridge has the third longest cantilever span in the world and was one of the last large cantilever bridges ever built. Cantilever bridges have increasingly been superseded by cable-stayed bridges, which are more economical.

The cantilever section of the 2½-mile-long Commodore Barry Bridge (1974) over the Delaware River in Pennsylvania is the longest cantilever bridge span in the United States. The main span is 1,644 feet long, with a clearance of 197 feet, but the cantilever structure forms part of a longer bridge built of truss beams. The downward forces in the cantilever arms that create the main span are balanced over the central piers and cause uplift forces in the anchor (or back) span. These forces are carried by the anchor arm and resisted by the outer piers and the weight of the back span.

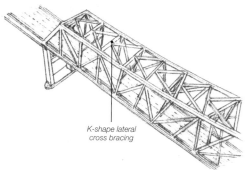

Lateral cross bracing (left)
Virtually all truss bridges rely on some form of cross bracing between their sides. Here, K-shape lateral cross bracing reinforces the upper part of each section. Similar cross bracing beneath the 79-foot-wide deck is strengthened by central longitudinal steel members.

K-shape lateral cross bracing

Subdivided Warren truss (right)
The subdivided Warren truss, used here for the approach roads and the main cantilever structure, is defined by the alternating direction of the diagonal members divided by the vertical members. These diagonal members transfer the loads back to the piers.

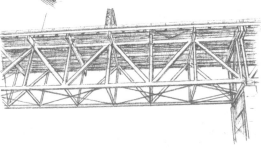

Subdivided Warren-type through truss

Under and through trusses (below)
The approach roads are simply supported subdivided Warren-type deck trusses (where the road deck sits on the truss beam) that switch to a through truss (where the road deck sits within the truss) at the cantilever section.

Subdivided Warren-type deck truss

Vejle Fjord Bridge

Expansion

The Vejle Fjord Bridge is divided into four sections to allow for the expansion of the prestressed-concrete box girders. Each expansion section is about 1,610 feet long.

The 5,617-foot-long bridge over the Vejle Fjord (1980) in Denmark has fifteen spans of approximately 360 feet. It was constructed using the free-cantilever method. Each double cantilever was constructed over the piers and built out to meet at the center span. The joint between each T is visible in the center of the span. In its permanent form, the bridge is a haunch-concrete box girder—the underside arcs from the support to the midspan—which transfers the load back to the piers.

Slender geometry (right)

The construction of a free cantilever requires the structure to have greater depth close to the supports getting shallower toward the outer ends of the cantilever arms, creating a slender section at midspan. The narrow piers support the haunch box-girder sections that arc from a depth of 20 feet over the piers to a shallow 8 feet in the center of the span. The maximum clearance is approximately 130 feet.

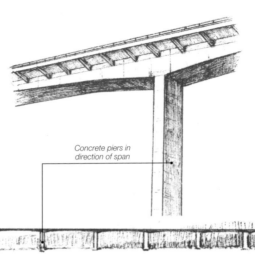

Concrete piers in
direction of span

Perpendicular piers (above)

To maintain the bridge's slim appearance, the concrete piers have been designed to be shallow in the direction of the span, which is made possible by the balanced cantilever form minimizing bending forces. The breadth of the piers offers lateral stability to the suspended deck.

Cantilever deck

The principal structural elements—the pier and central box girder—are the same width, and they support the wider deck. The deck is a cantilevered concrete slab that projects beyond the sides of the box girder on tapering prestressed transverse beams. The complete width of the thin concrete deck is 95 feet.

193

SUSPENSION BRIDGES

Introduction

The deck of a suspension bridge is hung from suspended cables in tension, which transfer the loads to anchorages at each end of the bridge or to towers, where they are carried to the ground. Traditional (or simple) suspension bridges comprise a deck between two suspended cables hung between two stationary points. Modern suspension bridges adopt the same principle but use two towers over which the cables are strung to create one main central span and two smaller side spans. The cables of a modern suspension bridge are typically made from a single strand wound continuously from one side of the bridge to the

other. At each pass, it is secured into anchorage blocks embedded deep in the ground. These anchorages are designed to resist the huge tensile forces in the cables that are generated by carrying the dynamic loads through the vertical hangers to the deck. Most of the weight, which combines the structure and the live loads traveling over it, is carried as tensile forces through the hangers and cables and is borne by the tower, which is placed under compression. The structural efficiency of suspension-bridge design makes it the most effective bridge system for crossing long distances with a single span.

Akashi Kaikyo
The longest suspension bridge in the world is the Akashi Kaikyo Bridge (1998), in Japan, which measures just under 2½ miles and has a main span of 1¼ miles.

SUSPENSION BRIDGES — Menai Straits Bridge

Rust prevention

The suspension chains of the Menai Straits Bridge were the longest ever constructed. Between fabrication and installation on-site, each link of iron chain was stored in warm linseed oil to prevent rusting.

Linking mainland Wales and the island of Anglesey across the treacherous waters of the Menai Straits is one of the earliest modern suspension bridges and, in its time, the longest span in the world. The 1,000-foot-long bridge has a 579-foot-span and stands 98 feet above sea level. Construction of the Menai Straits Bridge (1826) commenced in 1820 with the erection of the masonry piers; this was followed by the installation of the wrought-iron suspension chains supporting a double-roadway wooden deck that was laid over iron beams.

Masonry approaches (left)

Limestone piers support the approach roads and comprise four 52-foot-wide round arches on one side of the Straits and three on the other. The arched approaches culminate at the 150-foot-high marble-faced towers over which the suspension chains are hung.

Truss parapets (right)

The oak deck often succumbed to strong winds, and in 1893 it was replaced by a steel deck, increasing its dead weight from 698 tons to 1,120 tons. In 1938, the iron chains, rods, and steel deck were replaced by steel elements. Longitudinal truss parapets on the outer sides of the roadway support the deck and were concealed by cantilevered pedestrian walkways.

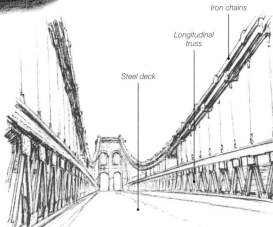

Iron chains

Longitudinal truss

Steel deck

Suspension chains (left)

The deck was originally supported by four sets of primary suspension chains secured to one another through eyebars at each end. Each primary chain weighed 133 tons and comprised 935 wrought-iron sections in four individual chains. These were hauled across the river by 150 workers and secured into 59-foot-deep anchorages tunneled into the bedrock and clamped in place with 10-foot-long iron bolts.

Clifton Suspension Bridge

Endurance

When the Clifton Suspension Bridge was built, the heaviest loads it was designed to carry were those of a horse and cart. The same structure continues to carry thousands of motorized vehicles over it every day.

The Clifton Suspension Bridge (1864) over the Avon River in Bristol, England, was the first commission for the young engineer Isambard Kingdom Brunel. Construction started in 1831, but the project was delayed by a number of problems. It was not completed until five years after Brunel's death. The total length of the bridge is 1,358 feet. The wrought-iron chains sit in saddles 13 feet below the top of the piers and descend 69 feet, giving the 31-foot-wide deck a 3-foot camber (a slight upward curve) and a clearance above the river of 249 feet.

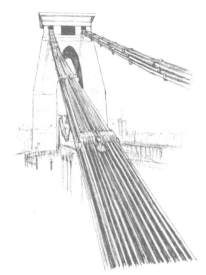

Iron links (left)

The suspension chains are made from wrought-iron links bolted together every few yards instead of wire bundled into cables, which later became standard. Each section of chain comprises ten or eleven flat iron links stacked in three independent vertical layers. Some of the chain links came from the dismantled Hungerford Bridge in London, also designed by Brunel.

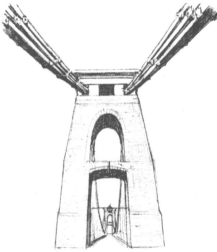

Suspension rods (below)

The main span of 702 feet comprises a deck suspended from eighty-one pairs of wrought-iron suspension rods. Ranging in length from 3 feet to 66 feet, these rods are arranged vertically and secured at the top end to the main chains and at the bottom end to the road deck.

Pylons (above)

The elliptically arched pylons, each 85 feet high, are constructed of Pennant stone and stand on sandstone abutments. Each pylon houses roller-mounted saddles over which the chains are hung, allowing for longitudinal movement in response to shifting loads and temperature variations, and to prevent transverse movement.

John A. Roebling Bridge

Suspension cables
The John A. Roebling Bridge had more than 10,000 wires woven between the piers to create the two main suspension cables supporting the deck with a clearance of 98 feet.

Posthumously renamed after its designer, the John A. Roebling Bridge over the Ohio River on the border between Kentucky and Ohio was the longest suspension bridge in the world when it opened to the public in 1866. With prevailing knowledge and expertise of suspension-bridge design, Roebling understood that the Ohio River was too wide for a suspension bridge to cross from shore to shore, so he designed a bridge with two masonry towers standing in the river 980 feet apart.

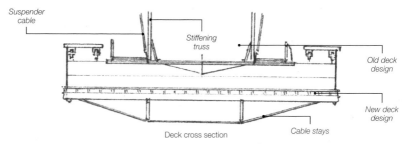

Suspender cable

Stiffening truss

Old deck design

New deck design

Deck cross section

Cable stays

Deck structure (above)

The deck is supported by iron suspender cables that hang from the main cables. Adjacent pairs of suspender cables are connected by wrought-iron beams, over which the deck is laid. Each deck section is strengthened between lateral beams by longitudinally arranged members and diagonal and transverse bracing, which provide lateral stiffness.

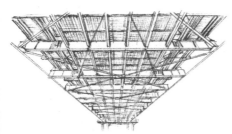

Widened deck

By the end of the 19th century, the deck had been widened and an extra pair of suspension cables had been laid over the towers to increase its load capacity. The original deck surface is relatively thin, with cable stays under the deck and longitudinal trusses for stiffness. The new 59-foot-wide deck is strengthened and stiffened by rigid and wire types of cross bracing.

Masonry piers

The foundations of the 75-foot-high limestone-faced sandstone piers sit on thirteen perpendicularly arranged layers of oak timbers bolted together and encased in concrete. Originally, the piers were designed to carry a much larger structure than was eventually built because of the disruption of the American Civil War.

Brooklyn Bridge

The Brooklyn Bridge (1883), in New York, was the longest suspension bridge in the world when completed. To span such a wide expanse of water, four suspension cables were used instead of two. These were strung over saddles in both masonry pylons above the iconic pair of pointed arches. Secondary cables radiating out from the pylons give the bridge its distinctive fanlike appearance, making it a hybrid of the cable-stayed and suspension systems for the portion close to the pylons.

Steel cables
The Brooklyn Bridge was the first suspension bridge to use steel-wire cables. Previously, suspension bridges used iron cables or chains, which were heavier and less structurally efficient.

Cable anchors

The four main cables were spun from 5,434 strands of continuously wound wire strung back and forth across saddles in the bridge, then tightly bound together and anchored at both shores. Each of the four cables could carry 12,300 tons. The vertical cables, or hangers, were manufactured from more than 14,300 miles of wires intertwined like a rope.

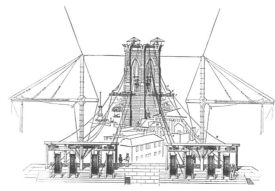

Overhead view

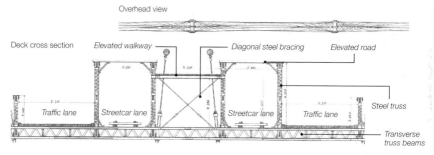

Deck cross section | Elevated walkway | Diagonal steel bracing | Elevated road

2,261 | *1,150* | *2,991*

Traffic lane | Streetcar lane | Streetcar lane | Traffic lane | Steel truss

Transverse truss beams

Deck section (above)

The deck originally carried two lanes of elevated railroad, two lanes for streetcars, and two lanes of vehicular traffic, with an elevated pedestrian walkway in the middle. Longitudinal steel trusses on the outside of the deck and larger ones separating the railroad from the outer lanes support the deck. The base of the deck comprises transverse truss beams spanning beneath the longitudinal trusses. The deck is strengthened with diagonal steel bracing to create lateral stability.

Elevation (below)

The entire length of the bridge is nearly 1¼ miles, with a main span between the pylons of 1,594 feet and a maximum clearance of 134 feet. The 276-foot-high pylons once towered over the New York skyline.

George Washington Bridge

Busiest bridge
More than five million vehicles crossed the George Washington Bridge in its first year. Today, nearly 100 million vehicles cross the bridge each year, making it the busiest bridge in the world.

The George Washington Bridge (1931) over the Hudson River in New York was revolutionary in its time, with a main span more than twice as long as any other in the world. It was constructed with just one deck but designed to support two. A second deck was later added and opened in 1962. The engineer, Othmar Ammann, knew that the weight of the deck could be used to stabilize the bridge, eliminating the need for stiffening trusses, which were common in earlier suspension bridges. Each of the bridge's 105-foot longitudinal deck beams weighs 73 tons. Four main steel suspension cables, each nearly 3 feet in diameter, carry the 118-foot-wide twin decks across the 3,500-foot span between towers. The length of the bridge between abutments is 4,760 feet, with a clearance of 213 feet.

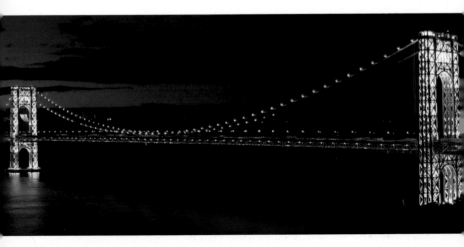

Additional deck

The additional lower deck was constructed beneath the existing deck using 26-foot-deep steel subdivided Warren-type trusses. The addition of a deck forms a structural extension because the top deck and lower deck are acting as a single structural entity instead of one being hung from the other.

Wire cables

Each of the four main cables comprises a single strand wound back and forth across the river sixty-one times and is secured in anchorage blocks that are made from 287,000 tons of concrete. Each strand is fabricated from 434 individual wires with a total length of 107,000 miles.

Trussed towers

Each of the two steel towers supporting the four suspension cables is 600 feet high and weighs more than 23,000 tons. The towers were originally intended to be covered in concrete and granite, but the appealing functional appearance of the exposed steel-truss skeleton resulted in them remaining uncovered, also saving money.

Golden Gate Bridge

Distinctive profile
The towers of the
Golden Gate Bridge,
which soar to a
height of 746 feet,
are made up of
hollow steel cells
riveted together.
After the towers
were completed, the
cables were spun
back and forth
across the gulf.

The iconic structure (1937) spanning the Golden Gate at
the mouth of San Francisco Bay as it opens out into the
Pacific Ocean is one of the world's most famous bridges.
The 89-foot-wide deck is suspended from 250 pairs of
suspender cables hanging from a pair of main cables. Each
main cable is 3 feet in diameter, 7,650 feet long, and
fabricated from 27,572 strands of galvanized-steel wire
with a total length of 80,000 miles. Securing these huge
cables and their tensile loads requires a concrete anchorage
weighing 66,000 tons at each end of the structure.

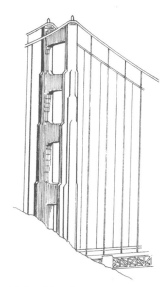

Environmental effects (below)

Temperature changes cause the deck to expand and contract, which affects the deck's length and, consequently, its height above water. The Golden Gate's vertical variation is as much as 16 feet, offering a maximum clearance of 220 feet. The deck is designed to reduce wind resistance and constructed using cross-braced steel subdivided Warren-type trusses, allowing for the wind to pass through with minimal resistance.

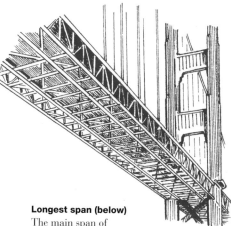

Styling (above)

The style of a bridge often reflects the era in which it was built. Few bridges are more evocative of their time than the Golden Gate Bridge, with its tapering Art Deco towers, whose horizontal bracing forms a Vierendeel truss. Each tower stands 746 feet tall and weighs 48,000 tons. The bridge's distinctive orange-vermillion color was chosen to complement the setting sun over the Pacific.

Longest span (below)

The main span of 4,200 feet broke all records when completed. The total length of the suspended structure, including the two 1,125-foot side spans, is 6,450 feet.

Lions Gate Bridge

Elevation
The 5,980-foot-long Lions Gate Bridge comprises four sections: a main span of 1,548 feet, two side spans of 613 feet, and a 2,200-foot approach road.

Originally opened as a two-lane road bridge in 1938, the Lions Gate Bridge near Vancouver, Canada, was unable to cope with the growing volume of traffic. It was decided to replace the deck while keeping the bridge open. The road deck was replaced in sections during nights and weekends between 2000 and 2001. Improvements in bridge design allowed for the existing steel towers and cables to carry the new deck which, although 35 percent wider than the original deck, weighed the same.

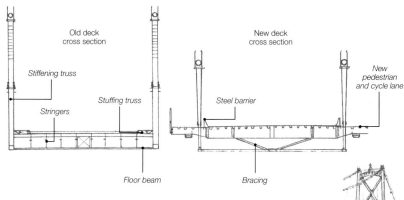

Old deck
cross section

Stiffening truss

Stringers

Stuffing truss

New deck
cross section

Steel barrier

New
pedestrian
and cycle lane

Floor beam

Bracing

Replacing the decks (above)

The original 40-foot-wide deck contained
two lanes of traffic and two pedestrian
walkways inside longitudinal trusses that
formed the sides of a U-shape structural
deck. The new 55-foot-wide deck had
a central section of 39 feet, from which
6½-foot-wide pedestrian and cycle lanes
are suspended on cantilevered sections on
each side. The trusses that once restricted
views from the deck were replaced by a
stiffening truss beneath the deck.

Tower trusses (right)

The towers are
strengthened by a
combination of lateral
supports—diagonal cross
bracing and narrower
tapered Vierendeel trusses
two-thirds up the tower
and also at its apex.
A horizontal beam
between the uprights
at the deck level also
supports the deck.

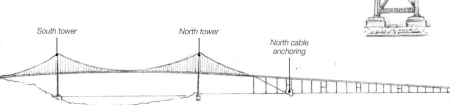

South tower

North tower

*North cable
anchoring*

Anchorage

The main cables are anchored differently
at each end. One is secured into the
bedrock where the deck meets the valley

side, and the other is saddled over the pier
separating the approach ramp from the
suspended section and taken down to an
anchorage on the valley floor.

Verrazano-Narrows Bridge

Expansion
Verrazano-Narrows Bridge's location exposes it to severe climatic conditions, causing it to close in winter or during high winds. Thermal expansion causes the deck elevation to change by 11½ feet between winter and summer.

The longest central span of any bridge in the country is the suspension bridge across the Narrows between Staten Island and Brooklyn. With a span of 4,258 feet, the Verrazano-Narrows Bridge was not only the longest bridge in the world when completed in 1964, but it also supported a double roadway. In order to carry the additional load, the bridge has two pairs of suspension cables instead of the more common single pair. The deck truss receives additional support by sitting on crossbeams between the uprights of each tower.

Double deck (left)

Fabricated from sixty steel-truss sections weighing 440 tons apiece, the double deck is supported from its upper surface. The lower deck is suspended below and comprises a cross-braced, 16-foot-deep subdivided Warren-type through truss, which provides lateral stability.

Nonparallel towers (right)

Standing 978 feet apart and 692 feet high, the design of the steel towers, which stand on 167-foot-deep foundations, had to accommodate the curvature of the Earth—their tops being $1\frac{2}{3}$ inches farther apart than their bases.

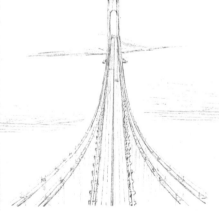

Twin cables

Each side of the bridge is supported by a pair of cables, each one made up of 26,108 strands and 143,000 miles long. The cables weigh 10,220 tons and measure almost 3 feet in diameter. The strands were laid in pairs by spinning wheels that passed back and forth between the anchorages, and they are clamped along the length of the cable.

Humber Bridge

Completed in 1981, the road bridge over the Humber River in Yorkshire, England, was once the longest suspension bridge in the world. The previous longest span of a suspension bridge with concrete towers was 1,995 feet, but the Humber Bridge's main span was 4,625 feet. The technological innovation that allowed for this significant increase was the novel use of hollow reinforced-concrete sections in the 508-foot-high towers. This allowed for the towers to be lighter and taller, reducing the load on the foundations.

Environmentally responsive
The decision to construct a suspension bridge was influenced by the environmental conditions in the Humber River, which are characterized by a shifting estuary bed. Shipping channels are, therefore, constantly changing and only a suspension bridge, which uses only two piers to create a large span, was suitable.

Hollow concrete towers (below)

The towers were constructed using the slip-forming technique, in which the concrete is poured in formwork that rises with the structure, a construction technique also used for some skyscraper cores. The two slightly tapered pylons in each tower are horizontally cross braced to form a Vierendeel truss for extra stability.

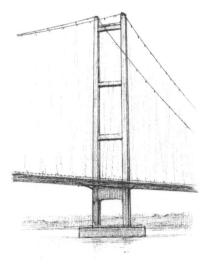

Steel cables (left)

Each of the two main steel cables is 27½ inches in diameter and comprises thirty-seven strands fabricated from 404 lengths of ⅕-inch-diameter high-tensile wire. Every strand is tied and anchored at each end of the bridge before being spun into cables. Each cable weighs 6,000 tons, and each reinforced-concrete anchorage weighs 330,000 tons.

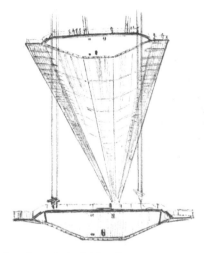

Aerodynamic deck (above)

The 92-foot-wide and 14¾-foot-deep deck is fabricated from 59-foot-long, 154-ton steel hollow box sections. These sections were fabricated off-site and suspended from hanger cables before being welded together. The materials and design reduce the deck's weight, and its razor-sharp aerodynamic profile increases its stability by minimizing the lateral forces created by wind loads.

Tsing Ma Bridge

Side spans

The cables over each of the Tsing Ma Bridge side spans are anchored at the ground instead of at deck level. One set is straight, not curved, because it carries no vertical load.

The world's longest suspension bridge carrying both rail and road traffic is the Tsing Ma Bridge (1997) in Hong Kong, China. The 7,087-foot-long bridge carries a double deck with six lanes of traffic above two railroad lines and two emergency traffic lanes. The asymmetrical structure comprises three sections: a main span, a side span partly suspended and partly supported on concrete piers, and an approach road supported on concrete piers.

Contained double deck (right)

The depth of the 54,000-ton double deck exposes it to high winds in an area prone to extreme weather. The deck has been designed to minimize wind resistance. Its aerodynamic-edge profile encourages crosswinds to pass around it, while a continuous opening along the middle of the upper deck prevents air pressure from building up within the deck.

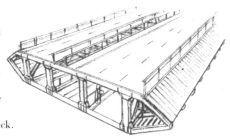

Bridge span (above)

The two main cables are strung over saddles at the top of the 676-foot-high and 57,000-ton reinforced-concrete towers. The side spans comprise a partially suspended section and an approach road supported on three evenly spaced reinforced-concrete piers 236 feet apart.

Anchorage blocks (below)

The two main cables are made from 33,400 steel wires each about ⅕ inch in diameter. Weighing a total of 29,400 tons, the two cables support a load of 117,000 tons and are secured at each end in anchorages weighing 220,000 and 275,000 tons, respectively. The face of these concrete structures is angled so that it is perpendicular to the line of the tensile forces in the cables.

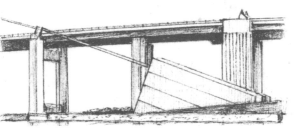

Great Belt Bridge

Suspended section
Suspension bridges are ideal in busy shipping areas because they give unobstructed clearances. The main span of the suspended section of the Great Belt Bridge provides a 5,328-foot-wide navigation channel.

The suspended section of the Great Belt Bridge (1998) forms part of an 8¾-mile-long parallel road-and-rail bridge (and tunnel) system across the strait between the Danish islands of Zealand and Funen. Most of the bridge comprises a continuous concrete beam supported on reinforced-concrete piers. An interesting feature of the bridge is that the suspension-cable anchorages are not land based. Massive concrete sea anchorages resist the tensile forces with profiled wedges providing a frictional resistance beneath the seabed.

Anchor cross section

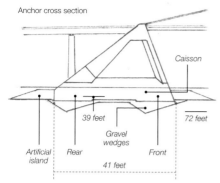

Caisson

39 feet   *72 feet*

Gravel wedges

Artificial island   *Rear*   *Front*

41 feet

Continuous suspension (below)

The deck is suspended continuously on two 10,100-foot-long cables between anchor points 1⅔ miles apart. This design feature causes the 833-foot-high reinforced-concrete towers to be higher than normal. The towers' tapered form reflects the compressive forces they are under, tapering from their solid bases and separating 69 feet above sea level into two pylons cross braced midway and at their top.

Sea anchorage (above)

The anchorages' hollow shape reduces the amount of concrete without compromising performance, making them lighter structures, aesthetically and physically. The shape reflects the tensile forces the anchorage is resisting. Each anchorage is 400 feet long at its base with a foundation 72 feet beneath the seabed and angled to resist the horizontal loads pulling it inward.

Deck section

The 102-foot-wide and 13-foot-deep aerodynamic welded steel box-girder deck is designed to resist wind loads. Transverse trusses inside the box section increase the deck's rigidity.

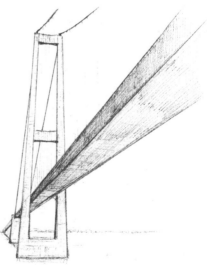

Deck cross section

Transverse truss

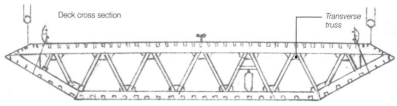

SUSPENSION BRIDGES

Akashi Kaikyo Bridge

Earthquake-proof
The 1995 Kobe earthquake moved the towers of the Akashi Kaikyo Bridge apart during construction, causing the main span to increase by 3 feet. Hinges in deck girders are designed to increase resistance to wind and earthquakes.

Crossing the Akashi Strait between Kobe and Iwaya in Japan is the longest span of any suspension bridge in the world. The 12,831-foot-long deck is suspended 318 feet above the sea by two huge steel cables over 3 feet in diameter, the 36,830 strands of which would wrap around the world seven times. Construction began in 1988 and took ten years to complete, consuming 200,000 tons of steel and 50,000 cubic feet of concrete.

World's longest span (below)

The cables are hung over a pair of 974-foot-high pylons that create three spans: a central span of 6,532 feet and two outer spans of 3,151 feet. Note how the hangers connecting the deck to the suspension cables are secured above each vertical truss member. If they were attached midway between vertical members, then each member would also take bending moments.

Concrete anchorage (above)

With such a large span generating considerable tensile forces, the cables have to be secured by huge anchorages. Each one is constructed using 386,000 tons of concrete and has been combined with the abutments. Note how the approach road is a concrete beam supported on the abutments, beyond which the deck becomes a subdivided Warren-type steel-truss box.

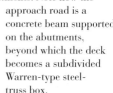

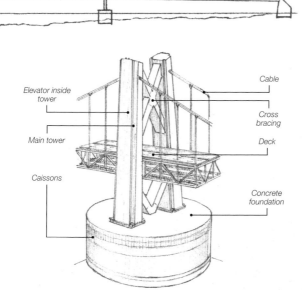

Elevator inside tower

Main tower

Caissons

Cable

Cross bracing

Deck

Concrete foundation

Foundations

The bases of each pylon are set into deep circular foundations constructed with concrete and steel to provide a solid base for the towers. Each tower is cross braced for lateral stability but remains relatively open to minimize wind forces.

Si Du River Bridge

Asymmetry
One side of the Si Du River Bridge comprises an approach road, whereas the deck on the other side meets the valley wall at the base of the tower, creating an asymmetrical profile.

Crossing the valley of the Si Du River in Hubei Province, China, at a height of approximately 1,640 feet, is the world's highest suspension bridge. The 4,478-foot-long bridge opened to traffic in November 2009. The bridge has a main span of 2,952 feet, but there are no suspended side spans because of the constricted geological setting. The main suspension cables are strung over the H-shape towers and then anchored directly into bedrock behind.

Deck cross section

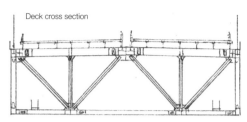

Truss deck (left)

The 85-foot-wide deck is
a frame made up of a series
of prefabricated steel trusses.
Longitudinal subdivided Warren-
type trusses run along the outside
of the entire length of the deck.
These are cross braced with regular
and evenly spaced, transverse
trusses that stiffen the structure.

Nonhorizontal deck (right)

The H-shape reinforced-concrete
pylons on each side of the valley
differ in height by 16 feet, creating
a slight gradient in the road deck.
Note how the main cables behind the
towers have no hangers because they
carry no vertical loads and serve only
to anchor the tensile forces in the
main cable. The cables, therefore,
form a straight line behind each
tower but a catenary curve
across the main span.

Rocket-launched cable

The bridge is supported by
two cables, which were spun
continuously back and forth
across the valley. The first pilot
cable, instead of being carried
down and across the valley, was
fired across by a rocket.

Introduction

Cable-stayed bridges are a type of suspension bridge, but they differ fundamentally from conventional suspension bridges. They share the same principle of using a series of tensile cables to transfer loads from the deck to a tower, which is placed under compression, but instead of a continuous cable, the cables of cable-stayed bridges are independent of one another and support deck sections extending out from the tower. This makes it easy to measure the forces in each cable and adjust its length as necessary. It also makes it easier to maintain the bridge because

individual cables can be replaced without compromising the bridge's overall strength. Cable-stayed bridges contain the loads within their structure, thus eliminating the need for an anchorage. This makes them particularly well suited to situations in which anchorages are not available, such as multiple spans or where the geological conditions provide an inadequate substrate. The deck load is transferred to the tower by symmetrical sets of cables, which balance the structure. Other types of cable-stayed bridges use inclined towers to balance the deck loads.

Rion–Antirion Bridge
Nearly 2 miles long, with a suspended deck of more than 1¼ miles, the Rion–Antirion Bridge in Greece is the longest multispan cable-stayed bridge in the world.

Skarnsund Bridge

High-strength concrete

The slender profile of the Skarnsund Bridge's deck and piers was achieved through recent advances in concrete technology that exploit the high-strength characteristics of modern concrete.

The Skarnsund Bridge (1991) in Norway was among the earliest examples of large cable-stayed bridges. With a total length of 3,314 feet, the bridge is a twin-pylon structure with three spans—one main span and two different-size outer spans. The main span is 1,740 feet long and 147 feet above the water. One side span is connected directly to approach roads on the cliff and the other is partially supported by four concrete piers that support the approach roads farther inland. The 43-foot-wide and 6½-foot-deep box-girder deck is the longest concrete cable-stayed deck in the world.

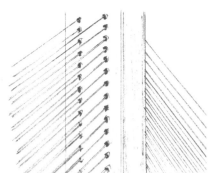

Cable anchorage (left)

Twenty-three pairs of cables weighing a total of 1,135 tons attach to each side of the mast at the apex of each tower. The cables vary in diameter from 2 inches to 3⅓ inches.

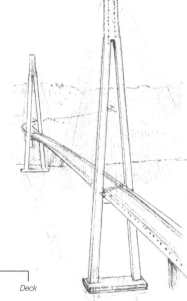

Triangular tower (right)

The A-frame shape of each 500-foot-tall tower is designed for its inherent lateral stability, being wide at its base and narrow at its top. The tensile forces are gathered at the top of the vertical mast and transferred down the two legs to the base of the tower and into the foundations.

Deck

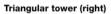

Flange

Crossbeam

Deck support

The load from the deck is supported both by the cables and the crossbeam between the legs of the tower. Small flanges on the side of the deck close to the tower are designed to prevent significant longitudinal movement in the deck. The deck's thin, aerodynamic triangular section makes it lightweight and structurally stable.

Pont de Normandie

Continuous beam
The approach ramp of the Pont de Normandie is made from a continuous concrete beam supported on piers that rise incrementally from the shore to the A-frame tower.

Cable-stayed bridge design has developed rapidly in recent years. A demonstration of this progress is the Pont de Normandie, which marked a revolution in cable-stayed-bridge design when constructed. When it opened in 1995, it was the world's longest cable-stayed bridge (7,030 feet) and had the world's longest cable-stayed span (2,808 feet), but the bridge has since been surpassed by many larger and longer cable-stayed bridges.

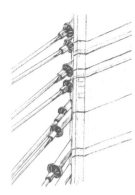

Cable connectors (left)

The deck is supported every 64 feet 4 inches by twenty-three pairs of steel cables on each side of the tower. These splay in a fanlike arrangement from the apex of the tower, where they are tied into steel sleeves embedded into a steel box, which is built into the mast. These transfer the tensile forces in the cables into compression forces in the towers.

Approach ramp

The last section of the approach ramp spans over the crossbeam in the tower and connects to the lighter steel box girder in the bridge's main span. The deck's slender profile is designed to minimize lateral forces generated by crosswinds.

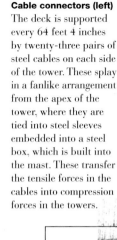

Approach ramp fabricated from a continuous concrete beam

A-frame tower (right)

The overall strength of the bridge derives from the structural efficiency of the A-shape frame of the boxed reinforced-concrete pylons, which transfers the loads from the deck to the ground through its legs. The two legs provide a solid base and the horizontal beam supports and laterally restrains the deck. The cables are secured at the vertical section of the mast above the point where the two legs join.

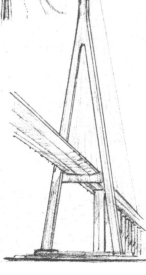

Vasco da Gama Bridge

Tied piers

The Vasco da Gama Bridge's legs taper outward below the crossbeam to the base and support the compressive forces from above. The legs are tied at their bases by concrete foundations that protect the piers from accidental impact from ships.

The 2,710-foot cable-stayed section of the Vasco da Gama Bridge (1998) forms a small part of the 10½-mile-long bridge across the Tagus River in Portugal. The main cable-stayed span is 1,377 feet long and the side spans are 666 feet long. The 98-foot-wide steel box-girder deck of the six-lane highway is carried across the three spans by 192 cables suspended from the two H-shape pylons. The inward-leaning lampposts on the deck are specially designed not to disturb the marine environment by shining directly on the water.

Deck support (left)
In the absence of any support beneath the deck, two cables on each side of the road deck on the inner face of the tower support the deck as it passes through the tower.

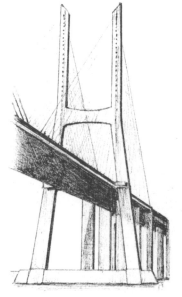

H-frame towers (right)
The cables are secured to the upper vertical portions of the 508-foot-tall H-frames and attached to the outer edge of the road deck. The tower structure is stiffened by the large, slightly haunched crossbeam above the road deck. Note there is no crossbeam at the head and none at deck level, meaning the deck is suspended throughout (the deck does, however, appear to be secured to the legs for lateral restraint). The foundations of the towers' legs form a protective crossbeam visible above the water level.

Twin cables
Some 184 cables of varying thickness radiate from the four vertical masts in a harplike arrangement. In a true harplike pattern, each cable stay is parallel to the next. Here, they are not quite parallel, but neither do they radiate from a single compressed point as in the case of a fanlike pattern.

Millau Viaduct

Piers

Each of the Millau Viaduct's piers tapers from 89 feet at the base to 33 feet at the deck, creating a form that reduces its presence in the landscape.

Bridging the wide valley created by the Tarn River in southern France, the Millau Viaduct (2004) is the longest cable-stayed bridge (1½ miles) and the tallest bridge (1,125 feet) in the world. The structure is divided into eight spans created by seven piers supporting 295-foot-high masts. Eleven pairs of cables on each mast support the 40,000-ton steel road deck, which has a gradient from north to south of 3 percent.

Longest cable-stayed deck (above)

The equally spaced piers range in height from 253 feet to 800 feet, creating six central spans of 1,122 feet and two outer spans of 669 feet. The cables are arranged in a radial formation attached at equally spaced anchorages on the road deck and up the pylons. The upper part of the pylon above the cable anchors serves no structural purpose.

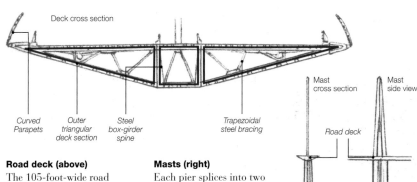

Deck cross section

Curved Parapets

Outer triangular deck section

Steel box-girder spine

Trapezoidal steel bracing

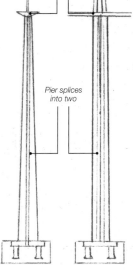

Mast cross section

Mast side view

Road deck

Pier splices into two

Road deck (above)

The 105-foot-wide road deck comprises three parts: a central spine fabricated from a 13-foot-wide and 13¾-foot-deep steel box girder; hollow outer deck sections fabricated from triangular elements welded to the spine; and trapezoidal steel bracing inside the deck to stiffen it. The cables connect to the central box-girder spine. The curved 10-foot-high parapets significantly reduce wind loads.

Masts (right)

Each pier splices into two beneath the road deck to increase the structure's longitudinal flexibility in order to accommodate the road deck's expansion and contraction. The 285-foot-high and 770-ton A-frame masts above the deck mirror this structural response and can be seen in the example of the elevations at right.

Charilaos Trikoupis (Rion–Antirion) Bridge

Alternative design
The distance of the strait favored a suspension bridge, but geological and seismic considerations forced an alternative design approach for the Rion–Antirion Bridge, leading to a multispan cable-stayed bridge.

The 9,449-foot-long Rion–Antirion Bridge (officially Charilaos Trikoupis Bridge, 2004) connects the Peloponnese to mainland Greece. It comprises approach roads and three main spans and two side spans, which are 1,837 feet and 938 feet long, respectively. Four pylons, each with two sets of cables comprising twenty-three pairs on both sides of the deck, support the 7,388-foot-long road deck. The deck is a composite cross-braced, steel-frame structure of longitudinal I beams transversely braced with steel beams and strengthened by concrete slabs between.

Elevation (above)

The cable-stayed bridge section is suspended from four towers. Expansion joints at each end of the suspended section as well as the deck's structural independence from the four piers (at deck level) allow for the deck to respond flexibly and safely to climatic and seismic conditions by accommodating longitudinal deck movements of up to 16 feet.

Foundations (below)

In the absence of bedrock at an appropriate foundation level, the 213-foot-tall pylons sit on huge radial reinforced-concrete foundations that are 295 feet in diameter, embedded in the soil, and supported on friction piles in the seabed.

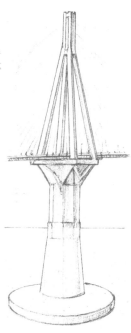

Square towers (above)

The reinforced-concrete towers are designed to transfer the loads from the deck directly and efficiently to the piers. The tensile forces in the cables are gathered in the mast and transferred down the four legs of each tower to the corner of the capital of each pier. The four-leg configuration provides additional stability in the event of an earthquake. The capital transfers the compressive loads into the octagonal piers and into the foundations.

Sutong Bridge

Triangular pylons

Sutong Bridge's inverted Y-shape legs are connected by posttensioned tie beams beneath the deck and designed to resist the impact of a 55,000-ton ship. The crossbeam ties the pylon legs and resists the outward-thrust forces.

The navigation channel in the 5-mile-long Sutong Bridge (2008) across the Yangtze River in China is created by the longest cable-stayed bridge in the world. Two sets of cable stays (one on each side of the deck) connect the pair of 980-foot-high distinctive Y-shape pylons to the steel box-girder deck to create a main span of 3,570 feet. With the bedrock lying at a depth of 885 feet beneath the river, the pylons and approach piers are supported in the overlying soils by hundreds of friction piles bored to a depth of around 330 feet.

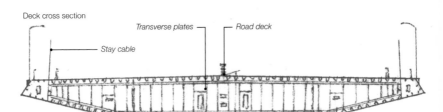

Deck cross section

Stay cable

Transverse plates

Road deck

Box-girder deck (above)

The cable-stayed deck section is a
134-foot-wide and 13-foot-deep steel
box girder fabricated from 52-foot-long
500-ton segments. Each segment was
hoisted from barges and welded together.
Longitudinal, closed steel troughs and
transverse plates every 13 feet along
the box section stiffen the deck,
with the transverse plates increasing
in frequency near the pylons.

Cable connection

The cables fan out from the upper
portion of the pylon, meeting the center
span at regular 52-foot intervals and
the back span at 39-foot intervals. The
deck was constructed in sections and
the individual cables were adjusted as
each deck section was laid. The longest
cable to the midspan is 1,893 feet long
and weighs 65 tons.

Approach

The approach roads are
fabricated from hollow,
single-cell concrete sections
forming a continuous girder
over transverse concrete piers.

Hangzhou Bay Bridge

Coastal exposure
Hangzhou Bay Bridge's potential exposure to typhoon winds, earthquakes, and extreme tides complicated both the design and construction process, which took more than a decade to complete.

The two cable-stayed sections of the 22-mile-long Hangzhou Bay Bridge (2008) in eastern China provide navigation channels in what was (until 2010) the longest seaborne bridge. The cable-stayed sections of the 121-foot-wide streamlined road deck are made from 49-foot-long and 11½-foot-deep steel box-girder sections. These are attached to approach roads constructed from concrete box-girder sections that form a continuous beam supported on pairs of concrete piers.

North and south sections (below and right)

The south channel comprises a two-span A-frame single pylon. The three-span north channel comprises a pair of diamond-shape pylons. The main spans of the north and south channels are 1,470 feet and 1,043 feet, respectively. Each pylon supports a double set of cables secured into anchor plates on adjacent sides of the road deck.

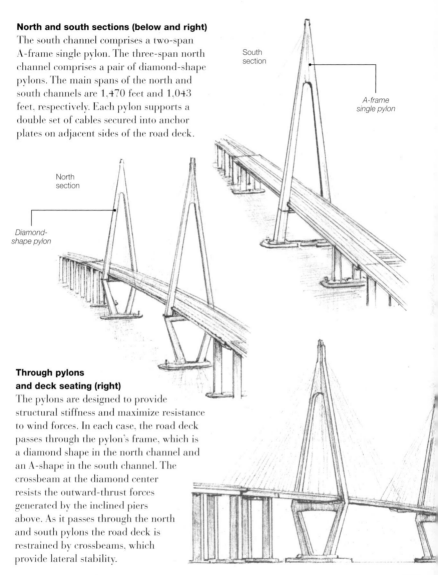

South
section

*A-frame
single pylon*

North
section

*Diamond-
shape pylon*

**Through pylons
and deck seating (right)**

The pylons are designed to provide structural stiffness and maximize resistance to wind forces. In each case, the road deck passes through the pylon's frame, which is a diamond shape in the north channel and an A-shape in the south channel. The crossbeam at the diamond center resists the outward-thrust forces generated by the inclined piers above. As it passes through the north and south pylons the road deck is restrained by crossbeams, which provide lateral stability.

Stonecutters Bridge

Crossing the busy Rambler Channel in Hong Kong, China, the Stonecutters Bridge (2009) comprises a 5,236-foot-long twin-pylon cable-stayed structure with a main span of 3,340 feet—the second longest cable-stayed span in the world. The two 978-foot-high needle towers support the main deck with twenty-seven pairs of cables extending from both sides of each needle tower. The cables are anchored to the main span at 59-foot intervals and to the side spans at 33-foot intervals.

Approach

The 167-foot-wide and 13-foot-deep twin deck separates as it approaches each tower with three lanes of traffic on each side. Strength is maintained by cross girders that tie the twin decks together. These can be seen between the decks in the midspan section.

Needle tower (below)

The tapering form of the needle towers from the broad base to the pointed apex reflects the structural forces as the pylon cantilevers up from its foundations. The first 574 feet is constructed from reinforced concrete, above which the concrete core is surrounded by a skin made of thirty-two stainless-steel sections. The composite design reduces oscillation caused by resonance in the tower and in the cables.

Cable connections (above)

The bottom three sets of cables are set directly into the concrete pylon core. All other cables are secured into steel anchor boxes. At the top of each cable, ⅔-inch-diameter stainless-steel shear connectors, each about 1 foot long, transfer the load between the concrete core, steel skin, and anchor box.

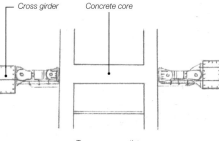

Cross girder   Concrete core

Tower cross section

Twin deck

The deck of each side span is fabricated from a continuous concrete box-girder beam supported on four concrete piers. The main span comprises a lightweight, aerodynamic steel box girder that minimizes dead loads. The deck is strengthened and stiffened internally by transverse and longitudinal fin plates. Strength is maintained by cross girders that tie the twin decks together.

Jiaozhou Bay (Qingdao Haiwan) Bridge

Movement joints
The different structural forms of the cable-stayed sections and the beam sections of the approach roads to the Jiaozhou Bay Bridge are clearly visible. They are separated by movement joints to allow for expansion and contraction between the structural elements.

The navigation channels in the 26⅓-mile-long bridge system (2011) in Jiaozhou Bay on China's east coast are created by three separate structures—one of which is suspended and two of which are cable-stayed. The suspended section comprises a single pylon supporting the twin road deck from its center, creating two spans. The shortest cable-stayed section comprises two adjacent H-shape pylons, each supporting one section of the twin road deck and creating two equal spans. The longest cable-stayed section comprises two sets of adjacent H-shape pylons creating one long main span and two shorter outer spans, which are also supported by concrete piers.

Paired pylons (right)

There are four vertical pylons across the width of the twin decks, and each half span of the bridge is supported by a total of twenty-four cables. Distributing the cables between four separate masts allows for them to be arranged in a harp design, in which they are parallel instead of fanning out from a single narrower point. It also increases the deck spans between suspension points. Each pair of vertical masts is stiffened beneath the road deck by a horizontal beam supporting the deck as well as a shared foundation at their base.

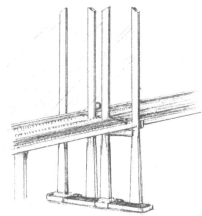

Twin deck (left)

The arrangement of the road into a twin deck allows for the loads to be divided into two, effectively creating two individual bridges. In the cable-stayed sections, this reduces the size of the pylons and the number and length of the cables. It also spreads the load more effectively into the foundations.

Bridge system (right)

The cable-stayed sections of the Jiaozhou Bay Bridge provide navigation channels in a vast bridge system that comprises the longest bridge over water in the world, including a major intersection in the middle of the bay, where three separate branches of the bridge converge.

Glossary

A-FRAME TRUSS a truss whose structure comprises two opposing diagonal members joined at their apex and by a horizontal member at their middles, resembling an A

ABUTMENT the foundation structure at each end of the bridge that supports the loads

ANCHOR ARM the section of a cantilever bridge that is secured to the anchorage

ANCHORAGE the point at which the bridge's supporting elements, such as suspension cables, are anchored to the ground

APEX the uppermost part of a bridge or part of its structure

APPROACH RAMP a section of the deck that provides the link between the road and the main bridge span

BACKSTAY CABLE a major cable on a cable-stayed bridge that is connected to the ground behind the mast or tower

BALUSTRADE a railing along the side of the bridge, usually comprising a series of upright supports that are topped by a horizontal rail

BASCULE a counterbalancing mechanism exploited in movable bridges in which the deck rotates about a pivot like a seesaw

BENDING MOMENT the sum of forces applied to a beam that cause it to bend. Bending moments create opposing tensile and compressive stresses over the cross section.

BOLT HOUSING the casing surrounding an aperture through which a bolt is designed to sit

BOWSTRING ARCH TRUSS also known as a tied arch, in which the lateral thrust forces in the line of the arch are taken by a lower chord, often doubling as the bridge deck. Unlike conventional arches, which apply vertical and horizontal thrust to their supports, tied arches apply only vertical thrust.

BOX GIRDER a hollow beam

BRACE a supporting member that provides additional strength or rigidity in a structure

BULKHEAD a partition in a structure that can also strengthen it

CABLE STAY one of a number of independent high-tensile wires extending from a mast or tower to support the bridge deck. Collectively, the cables are arranged either by radiating from the top of the mast, in a fanlike pattern, or in parallel.

CAISSON a pressurized watertight and airtight chamber used in the construction of bridge foundations situated underwater

CANTILEVER a structure projecting in one direction from an upright support

CANTILEVER ARM the portion of a cantilever bridge that forms part of the main span

CATENARY CURVE the natural curve created by a suspended cable supported at both ends

CHORD the top and bottom longitudinal members in a truss

CLAPPER BRIDGE a primitive method of bridge construction using only flat stone slabs supported on stone piers

COFFERDAM a watertight enclosure placed onto the riverbed and drained to permit the construction or repair of a structure. *See also* **CAISSON**

COMPRESSIVE STRENGTH a material's ability to resist compressive forces

CORBEL a structural element that extends beyond a wall's surface to support members above

COR-TEN STEEL a patented type of weathering steel that does not require painting and forms a stable rustlike appearance

CREEP the gradual movement of a material or structure caused by its own mass or applied long-term load or in response to external factors

CROSSBEAM a transverse beam that connects two principal sides of a structure

CROSS BRACE a form of transverse bracing in a structure to increase strength and rigidity

CROWN the uppermost part of a structure, such as an arch

COUNTERWEIGHT a weight that acts as a counterbalance

DECK the main surface of a bridge over which traffic passes

DOVETAIL JOINT a type of joint comprising an interlocking mortise and tenon

DYNAMIC FORCE a force that induces acceleration or vibration in a structure

EXPANSION JOINT the space between two parts of a structure to allow for expansion and contraction

Glossary

EXTRADOS/EXTRADOSED the exterior curve of an arch

EXTRUSION the means by which a component (such as a steel wire) is created with a consistent and continuous section

EYEBAR a structural member comprising a solid bar with holes at each end, forming part of a chain

FALSEWORK a temporary structure, such as scaffolding, erected in the course of constructing a permanent structure

FERROCONCRETE a form of reinforced concrete wherein iron or steel bars are cast in concrete to increase the material's tensile strength

FIN PLATE a sheet of metal welded to a structure to increase strength and rigidity

FLANGE a section of a structural member that is widened for additional strength

FORMWORK a mold, usually made of wooden boards, into which concrete is poured

FRICTION PILE a pile used in foundations whose effectiveness depends on friction with the surrounding soil instead of end bearing

GIRDER A large beam. Girders are commonly associated with iron or steel but could be fabricated in any material, such as concrete.

GONDOLA the movable compartment or cabin suspended from cables on a transporter bridge

H-FRAME TOWER a structural form characterized by two vertical members joined together at midheight on a horizontal beam, forming an H shape

HANGER *see* **SUSPENDER CABLE**

HAUNCH BOX GIRDER a box girder with an underside that arcs from the support to the midspan. The deepening of the girder over the supports reflects the form of construction and its structural performance.

HELIX a cylindrical spiral or corkscrew structure

HOGGING MOMENT a bending moment that generates tensile forces in the top of a beam, causing the bottom surface to be concave. Opposite of sagging.

HYDRAULIC a mechanism that is operated by the movement of water under pressure

JOINT *see* **EXPANSION JOINT**

K-TRUSS a truss with two diagonal members between each pair of vertical supports. The diagonal elements, like those on the letter K, begin at the midpoint of one vertical and finish at the top and bottom ends of the adjacent vertical.

KEYSTONE the central stone at the apex of an arch

LATERAL STABILITY providing structural stability through the bridge's depth instead of along its length. *See also* **LONGITUDINAL**

LEAF the portion of a bascule bridge deck that is raised and lowered

LENTICULAR TRUSS a type of truss comprising two curved chords. The overall outline resembles a lens, from which it derives its name

LOAD the weight imposed on a bridge. Dead loads are those constant and unvarying forces the structure has to withstand, such as the bridge's inherent weight. Live loads are the unpredictable and varying forces caused by environmental factors or the changing weight of traffic over the bridge.

LONGITUDINAL of or pertaining to the length of the bridge

MAST a principal structural member projecting upward, often vertically, from the bridge

MEMBER a part or element of a structure

MORTISE-AND-TENON JOINT a simple interlocking joint originating in carpentry where two pieces of wood are joined securely by inserting part of the "male" component (tenon) into a corresponding hole in the "female" component (mortise)

PARAPET a low wall that forms a barrier along the outer edge of the bridge

PERPENDICULAR at right angles to a line or plane

PIER vertical elements of a bridge that support horizontal elements, such as a beam or truss

PILE a long vertical section of wood or concrete (usually cylindrical in section) embedded in the ground that forms parts of a structure's foundations

Glossary

PIN a bolt or similar component that joins two members together by being inserted through apertures in both and held securely

POSTTENSIONED CONCRETE a form of reinforced concrete in which spanning capabilities are enhanced by inducing compression forces in the parts of the structure likely to develop tensile forces. The method involves tensioning steel cables (tendons) within casing sleeves after the concrete has set.

PRATT TRUSS a type of truss in which the diagonal members on each side of the truss are orientated in the same direction and meet at the middle. In a simply supported configuration these diagonals work in tension.

PRECAST a concrete component that is manufactured off-site

PYLON a tall structural member that acts as a principal supporting element in a bridge

RADIAL spreading out from a common center, such as the spokes of a wheel

REINFORCED CONCRETE concrete that contains reinforcement bars or fibers to enhance its tensile capacity

RIB a small member used to stiffen or strengthen part of a structure

SADDLE the mechanism at the top of suspension-bridge towers over which suspension cables are laid. Their purpose is to allow for movement in the cables in response to climatic conditions and functional requirements.

SAGGING MOMENT a bending moment that generates tensile forces in the bottom of a beam, causing the top surface to be concave. Opposite of hogging.

SEGMENTAL ARCH a shallow type of arch that is formed by an arc of less than a semicircle

SEISMIC of or pertaining to earthquakes or the movement of the earth

SHEAR FORCE a force acting perpendicularly to the longitudinal axis of a beam

SICKLE ARCH an arch that resembles the shape of sickle, often being thicker at its midsection than at its ends

SLIP-FORMING a construction technique in which concrete is poured into a continuously moving mold or form. The technique is commonly used in the construction of tall structures with a consistent section, such as the core of a skyscraper or the pylons of cable-stayed or suspension bridges.

SPAN the distance between two supports or piers of a bridge

SPANDREL the area, often triangular in shape, between the horizontal deck, the curve of the arch (extrados), and the vertical abutment

SUSPENDER CABLE the cables, often referred to as hangers, that attach the road deck to the main suspension cables on a suspension bridge

TENSILE of or pertaining to a pulling force, such as tension

TENSION *see* **TENSILE**

THROUGH TRUSS a truss-type bridge in which the deck passes through the truss, usually on the lower chord

THRUST a pushing force acting on or exerted by a structure

TIED ARCHES *see* **BOWSTRING TRUSS**

TORSION a twisting force

TRANSVERSE BRACING strengthening that occurs across a structure instead of along it

TRUSS a conventional truss is a structure that derives its strength from the triangle. Each member is subject to compression or tension (or both, though not at the same time).

VAULTING a type of roofed structure formed by a series of arches or vaults

VIADUCT a type of bridge over land formed by a series of small (usually arched) spans

VIERENDEEL TRUSS a type of truss formed of rectangular (typically) instead of triangular sections and which therefore relies on the strength of its component members in bending as well as pure tension and compression

VOUSSOIR one of the wedge-shape pieces that makes up an arch or vault

WARREN TRUSS a type of truss made up of only horizontal and diagonal members. Conventional Warren-type trusses have no vertical members, so the alternating diagonal members form a series of equilateral triangles between the horizontal members. Warren trusses with vertical members inserted are called subdivided Warren trusses.

Resources

Books

*Bridge Engineering:
A Global Perspective*
LEONARDO FERNÁNDEZ TROYANO
(Thomas Telford, 2003)

Bridges: Aesthetics and Design
FRITZ LEONHARDT
(Deutsche Verlags-Anstalt
GmbH, 1983)

*Bridges: An Easy-Read Modern
Wonders Book*
CASS R. SANDAK
(F. Watts, 1983)

*Bridges: The Science and Art
of the World's Most Inspiring
Structures* DAVID BLOCKLEY
(Oxford University Press,
2010)

*Bridges: The Spans of North
America*
DAVID PLOWDEN
(W. W. Norton & Company,
2001)

*Bridges: Three Thousand Years
of Defying Nature*
DAVID J. BROWN
(Firefly Books, 2005)

Bridges of the World
TIM LOCKE
(Automobile Association, 2008)

*Bridges of the World: Their
Design and Construction*
CHARLES S. WHITNEY
(Dover Publications Inc., 2003)

*Bridges That Changed
the World*
BERNHARD GRAF
(Prestel Publishing Ltd, 2005)

*Brunel: The Man Who
Built the World*
STEVEN BRINDLE,
DAN CRUICKSHANK
(Phoenix, 2006)

*Creation of Bridges: From
Vision to Reality—The Ultimate
Challenge of Architecture,
Design, and Distance*
DAVID BENNETT
(Diane Pub Co, 1999)

*Dan Cruickshank's Bridges:
Heroic Designs That
Changed the World*
DAN CRUICKSHANK
(Collins, 2010)

*Handbook of International
Bridge Engineering and Design*
WAI-FAH CHEN, LIAN DUAN
(CRC Press, 2011)

*History of the Modern
Suspension Bridge: Solving
the Dilemma Between Economy
and Stiffness*
TADAKI KAWADA
(American Society of Civil
Engineers, 2010)

Integral Bridges
GEORGE L. ENGLAND,
NEIL C. M. TSANG, DAVID I. BUSH
(Thomas Telford, 2000)

*Landmarks on the Iron Road:
Two Centuries of North
American Railroad Engineering*
WILLIAM D. MIDDLETON
(Indiana University Press,
1999)

*Masterpieces: Bridge
Architecture & Design*
CHRIS VAN UFFELEN
(Braun Publishing AG, 2009)

*Structures: or Why Things
Don't Fall Down*
J. E. GORDON
(DaCapo Press, 2003)

*Superstructures: The World's
Greatest Modern Structures*
N. PARKYN
(Merrell Publishers Ltd, 2004)

Victorian Engineering
L. T. C. ROLT
(The History Press Ltd, 2007)

*What is a Bridge?: The Making
of Calatrava's Bridge in Seville*
SPIRO N. POLLALIS
(MIT Press, 2002)

Web sites

Bridge Hunter
www.bridgehunter.com
Web site database of historic or notable bridges
in the United States, past and present

Bridge Pix
www.bridgepix.com
Web site that includes an ever-expanding
database with more than 13,000 bridge
photographs

Bridge Pros
www.bridgepros.com
Web site dedicated to the engineering,
history, and construction of bridges

Bridges
www.brantacan.co.uk/bridges.htm
Basic Web site that examines the simple and
complex structures of bridges, providing
accurate and informative details. Text-based
with illustrations

Engineering Timelines
**www.engineering-timelines.com/
timelines.asp**
Web site dedicated to celebrating the lives
and works of the engineers who have shaped
the British Isles

Highest Bridges
highestbridges.com
Web site detailing in-depth knowledge of
500 of the world's highest bridges. Includes
plans, photographs, and illustrations.

Historic Bridges
www.historicbridges.org
Web site of photo-documented information
on all types of historic bridges in the United
States, parts of Canada, and the UK

Nicolas Janburg's Structurae
en.structurae.de
Web site offering information on works of
structural engineering, architecture, and
construction throughout history and from
around the world

Swiss Timber Bridges
www.swiss-timber-bridges.ch
Web site dedicated to detailing all the
wooden bridges of Switzerland; the site
currently contains 1,412 bridges documented
by 19,947 images and documents

U.S. Department of Transportation;
Bridge Technology
www.fhwa.dot.gov/bridge
Government Web site containing updated details
on bridge technology and bridge studies

Index

Index

Acknowledgments

AUTHOR ACKNOWLEDGMENTS

Edward Denison
I would like to thank all the staff at the British Library for their dedication to research and also Ian for his expert advice and direction throughout this project, without which this publication would not have been possible.

Ian Stewart
I would like to thank my wife Louise for all of her love, support, and encouragement as always. I would also like to thank Edward for the enjoyable and creative experience of writing together.

We are indebted to our editor, Caroline Earle, for her patience and professionalism and to everyone at Ivy Press for their invaluable contributions, including Kate Shanahan, Jamie Pumfrey, Adam Hook, and Jane and Chris Lanaway.